Statistics and Computing

Series Editors:
J. Chambers
W. Eddy
W. Härdle
S. Sheather
L. Tierney

Statistics and Computing

Peter Dalgaard

Introductory Statistics with R

With 48 Illustrations

 Springer

Peter Dalgaard
Department of Biostatistics
University of Copenhagen
Blegdamsvej 3
2200 Copenhagen N.
Denmark
p.dalgaard@biostat.ku.dk

Series Editors:

J. Chambers
Bell Labs, Lucent Technologies
600 Mountain Avenue
Murray Hill, NJ 07974
USA

W. Eddy
Department of Statistics
Carnegie Mellon University
Pittsburgh, PA 15213
USA

W. Härdle
Institut für Statisik und
 Ökonometrie
Humbolt-Universität zu Berlin
Spandauer Str. 1
D-10178 Berlin
Germany

S. Sheather
Australian Graduate School
 of Management
University of New South Wales
Sydney, NSW 2052
Australia

L. Tierney
School of Statistics
University of Minnesota
Vincent Hall
Minneapolis, MN 55455
USA

Library of Congress Cataloging-in-Publication Data
Dalgaard, Peter.
    Introductory statistics with R/ Peter Dalgaard.
      p. cm. — (Statistics and computing)
    Includes bibliographical references and index.
    ISBN 0-387-95475-9 (softcover : alk. paper)
      1. Statistics—Data processing.   2. R (Computer program language)   I. Title.   II. Series.
    QA276.4.D33   2002
    519.5—dc21   2002020947

ISBN-10: 0-387-95475-9   Printed on acid-free paper.
ISBN-13: 978-0387-95475-2

Printed in the United States of America.   (MVY)

9 8 7

springer.com

To Grete. For putting up with me for so long.

Preface

R is a statistical computer program, made available through the Internet under the General Public License (GPL). That is, it is supplied with a license that allows you to use it freely, distribute it, or even sell it, as long as the receiver has the same rights and the source code is freely available. It exists for Microsoft Windows 95 or later, for a variety of Unix and Linux platforms, and for the Apple Macintosh (OS versions newer than 8.6).

R provides an environment in which you can perform statistical analysis and produce graphics. It is actually a complete programming language, although that is only marginally described in this book. Here we content ourselves with learning the elementary concepts and seeing a number of cookbook examples.

R is designed in such a way that it is always possible to do further computations on the results of a statistical procedure. Furthermore, the design for graphical presentation of data allows both no-nonsense methods, for example `plot(x,y)`, and the possibility of fine-grained control of the output appearance. The fact that R is based on a formal computer language gives it tremendous flexibility. Other systems present simpler interfaces in terms of menus and forms, but often the apparent user-friendliness turns into a hindrance in the longer run. Although elementary statistics is often presented as a collection of fixed procedures, analysis of moderately complex data requires ad-hoc statistical model building, which makes the added flexibility of R highly desirable.

R owes its name to typical Internet humour. You may be familiar with the programming language C (whose name is a story in itself). Inspired by this, Becker and Chambers chose in the early 1980s to call their newly developed statistical programming language S. This language was further developed into the commercial product S-PLUS, which by the end of the decade was in widespread use among statisticians of all kinds. Ross Ihaka and Robert Gentleman from the University of Auckland, New Zealand, chose to write a reduced version of S for teaching purposes, and what was more natural than choosing the immediately preceding letter? Ross' and Robert's initials may also have played a role.

In 1995 Martin Maechler persuaded Ross and Robert to release the source code for R under the GPL. This coincided with the upsurge in open source software spurred by the Linux system. R soon turned out to fill a gap for people like me who intended to use Linux for statistical computing but had no statistical package available at the time. A mailing list was set up for the communication of bug reports and discussions of the development of R.

In August 1997 I was invited to join an extended international core team whose members collaborate via the Internet and which has controlled the development of R since then. The core team was subsequently expanded several times and currently includes 15 members. On February 29, 2000, version 1.0.0 was released. As of this writing, the current version is 1.5.0.

R implements a dialect of the S language. There are some differences, but in everyday use the two are very similar. However, some functions do differ, often because the R version tries to simplify things for the user. The differences are not all that big, but it would be silly not to take advantage of the improvements in a book at this level, so although the book might be used as an introduction to S-PLUS as well as R, the reader is urged to use R while working through it.

The book is based upon a set of notes developed for the course in Basic Statistics for Health Researchers at the Faculty of Health Sciences of the University of Copenhagen. This course has a primary target of students for the Ph.D. degree in medicine. However, the material has been substantially revised and I hope that it will be useful for a larger audience, although some biostatistical bias remains, particularly in the choice of examples.

This book is not a manual for R. The idea is to introduce a number of basic concepts and techniques that should allow the reader to get started with practical statistics.

In terms of the practical methods, the book covers a reasonable curriculum for first-year students of theoretical statistics as well as for engineering students. These groups will eventually need to go further and study more complex models as well as general techniques involving actual programming in the R language.

For fields where elementary statistics is taught mainly as a tool, the book goes somewhat further than what is commonly taught at the undergraduate level. Multiple regression methods or analysis of multifactorial experiments are rarely taught at that level but may quickly become essential for practical research. I have collected the simpler methods near the beginning to make the book readable also at the elementary level. However, in order to keep technical material together, Chapter 1 does include material that some readers will want to skip.

The book is thus intended to be useful for several groups, but I will not pretend that it can stand alone for any of them. I have included brief theoretical sections in connection with the various methods, but more than as teaching material, these should serve as reminders or perhaps as appetizers for readers who are new to the world of statistics.

Acknowledgements

Obviously, this book would not have been possible without the efforts of my friends and colleagues on the R Core Team, the authors of contributed packages, and many of the correspondents of the e-mail discussion lists.

I'm deeply grateful for the support of my colleagues and co-teachers Lene Theil Skovgaard, Bendix Carstensen, Birthe Lykke Thomsen, Helle Rootzen, Claus Ekstrøm, and Thomas Scheike, as well as the feedback from multiple students enrolled in the course in basic statistics for health science researchers. In addition, several people, including Bill Venables, Brian Ripley, and David James, have offered valuable advice on the manuscript.

Finally, profound thanks are due to the free software community at large. The R project would not have been possible without their effort. For the typesetting of this book TEX, LATEX, and the consolidating efforts of the LATEX2e project have been indispensable.

Peter Dalgaard
Copenhagen
May 2002

Contents

1

Basics

The purpose of this chapter is to get you started using R. It is assumed that you have a working installation of the software and of the ISwR package that contains the data sets for this book. Instructions for obtaining and installing the software are given in Appendix A.

The text that follows describes R version 1.5.0. As of this writing that is the latest version of R. As far as possible I present the issues in a way that is independent of the operating system in use and assume that the reader has the elementary operational knowledge to select from menus, move windows around, etc. I do, however, make exceptions where I am aware of specific difficulties with a particular platform or specific features of it.

1.1 First steps

This section gives an introduction to the R computing environment and walks you through its most basic features.

Starting R is straightforward, but the method will depend on your computing platform. You will be able to launch it from a system menu, by double-clicking an icon, or by entering the command "R" at the system command line. This will either produce a console window or cause R to start up as an interactive program in the current terminal window. In either case, R works fundamentally by the question-and-answer model: You

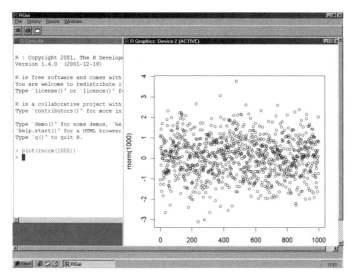

Figure 1.1. Screen dump of R for Windows.

enter a line with a command and press Enter (↩). Then the program does something, prints the result if relevant, and asks for more input. When R is ready for input, it prints out its prompt, a ">". It is possible to use R as a text-only application, and also in batch mode, but for the purposes of this chapter, I assume that you are sitting at a graphical workstation.

All the examples in this book should run if you type them in exactly as printed, *provided* that you have the ISwR package not only installed but also loaded into your current search path. This is done by entering

```
> library(ISwR)
```

at the command prompt. You do not need to understand what the command does at this point. It is explained in Section 1.5.3.

For a first impression of what R can do, try typing the following:

```
> plot(rnorm(1000))
```

This command draws 1000 numbers at random from the normal distribution (rnorm = *r*andom *norm*al) and plots them in a pop-up graphics window. The result on a Windows machine can be seen on Figure 1.1.

Of course, you are not expected at this point to guess that you would obtain this result in that particular way. The example is chosen because it brings several components of the user interface in action. Before the style of commands will fall naturally, it is necessary to introduce some concepts and conventions through simpler examples.

Under Windows, the graphics window will have taken the keyboard focus at this point. Click on the console to make it accept further commands.

1.1.1 An overgrown calculator

One of the simplest possible tasks in R is to enter an arithmetic expression and receive a result. (The second line is the answer from the machine.)

```
> 2 + 2
[1] 4
```

So the machine knows that 2 and 2 makes 4. Of course, it also knows how to do other standard calculations. For instance, here is how to compute e^{-2}:

```
> exp(-2)
[1] 0.1353353
```

The [1] in front of the result is part of R's way of printing numbers and vectors. It is not useful here, but it becomes so when the result is a longer vector. The number in brackets is the index of the first number on that line. Consider the case of generating 15 random numbers from a normal distribution:

```
> rnorm(15)
 [1] -0.18326112 -0.59753287 -0.67017905  0.16075723  1.28199575
 [6]  0.07976977  0.13683303  0.77155246  0.85986694 -1.01506772
[11] -0.49448567  0.52433026  1.07732656  1.09748097 -1.09318582
```

Here, for example, the [6] indicates that 0.07976977 is the sixth element in the vector. (For typographical reasons, the examples in this book are made with a shortened line width. If you try it on your own machine, you will see the values printed with six numbers per line rather than five. The numbers themselves will also be different since random number generation is involved.)

1.1.2 Assignments

Even on a calculator, you will quickly need some way to store intermediate results, so that you don't have to key them in over and over again. R, like other computer languages, has *symbolic variables*, that is names that can be used to represent values. To assign the value 2 to the variable x, you can enter

```
> x <- 2
```

The two characters <- should be read as a single symbol: an arrow point-
ing to the variable to which the value is assigned. This is known as the
assignment operator. Spacing around operators is generally disregarded
by R, but notice that adding a space in the middle of a <- changes the
meaning to "less than" followed by "minus" (conversely, omitting the
space when comparing a variable to a negative number has unexpected
consequences!).

There is no immediately visible result, but from now on, x has the value 2
and can be used in subsequent arithmetic expressions.

```
> x
[1] 2
> x + x
[1] 4
```

Names of variables can be chosen quite freely in R. They can be built from
letters, digits, and the period (dot) symbol. There is, however, the limita-
tion that the name must not start with a digit or a period followed by a
digit. Names that start with a period are special and should be avoided.
A typical variable name could be height.1yr, which might be used to
describe the height of a child at the age of 1 year. Names are case-sensitive:
WT and wt do not refer to the same variable.

Some names are already used by the system. This can cause some confusion if you use them for other purposes. The worst cases are the
single-letter names c, q, t, C, D, F, I, and T, but there are also diff, df,
and pt, for example. In practice, you get used to avoiding them after a
few accidents.

1.1.3   Vectorized arithmetic

You cannot do much statistics on single numbers! Rather, you will look at
data from a group of patients, for example. One strength of R is that it can
handle entire *data vectors* as single objects. A data vector is simply an array
of numbers, and a vector variable can be constructed like this:

```
> weight <- c(60, 72, 57, 90, 95, 72)
> weight
[1] 60 72 57 90 95 72
```

The construct c(...) is used to define vectors. The numbers are made
up but might represent the weights (in kg) of a group of normal men.

This is neither the only way to enter data vectors into R, nor is it generally the preferred method, but short vectors are used for many other
purposes, and the c(...) construct is used extensively. In Section 1.6

we discuss alternative techniques for reading data. For now, we stick to a single method.

You can do calculations with vectors just like ordinary numbers, as long as they are of the same length. Suppose that we also have the heights that correspond to the weights above. The body mass index (BMI) is defined for each person as the weight in kg divided by the square of the height in meters. This could be calculated as follows:

```
> height <- c(1.75, 1.80, 1.65, 1.90, 1.74, 1.91)
> bmi <- weight/height^2
> bmi
[1] 19.59184 22.22222 20.93664 24.93075 31.37799 19.73630
```

Notice that the operation is carried out elementwise, that is, the first value of bmi is $60/1.75^2$ and so forth and that the ^ operator is used for raising a value to a power. (On some keyboards, ^ is a "dead key" and you will have to press the spacebar afterwards to make it show.)

It is in fact possible to perform arithmetic operations on vectors of different length. We already used that when we calculated the height^2 part above since 2 has length 1. In such cases, the shorter vector is *recycled*. This is mostly used with vectors of length 1 (scalars) but sometimes also in other cases where a repeating pattern is desired. A warning is issued if the longer vector is not a multiple of the shorter in length.

These conventions for vectorized calculations make it very easy to specify typical statistical calculations. Consider, for instance, the calculation of the mean and standard deviation of the weight variable.

First calculate the mean, $\bar{x} = \sum x_i/n$:

```
> sum(weight)
[1] 446
> sum(weight)/length(weight)
[1] 74.33333
```

Then save the mean in a variable xbar and proceed with the calculation of SD = $\sqrt{(\sum(x_i - \bar{x})^2)/(n-1)}$. We do this in steps to see the individual components. The deviations from the mean are

```
> xbar <- sum(weight)/length(weight)
> weight - xbar
[1] -14.333333  -2.333333 -17.333333  15.666667  20.666667
[6]  -2.333333
```

Notice how xbar, which has length 1, is recycled and subtracted from each element of weight. The squared deviations will be

```
> (weight - xbar)^2
```

```
[1] 205.444444    5.444444 300.444444 245.444444 427.111111
[6]   5.444444
```

Since this command is quite similar to the one before it, it is convenient
to enter it by editing the previous command. On most systems running R,
the previous command can be recalled with the up-arrow key.

The sum of squared deviations is similarly obtained with

```
> sum((weight - xbar)^2)
[1] 1189.333
```

and all in all the standard deviation becomes

```
> sqrt(sum((weight - xbar)^2)/(length(weight) - 1))
[1] 15.42293
```

Of course, since R is a statistical program, such calculations are already
built into the program, and you get the same results just by entering

```
> mean(weight)
[1] 74.33333
> sd(weight)
[1] 15.42293
```

1.1.4 Standard procedures

As a slightly more complicated example of what R can do, consider the
following: The rule of thumb is that the BMI for a normal-weight individual should be between 20 and 25, and we want to know if our data
deviate systematically from that. You might use a one-sample t test to assess whether the 6 persons' BMI can be assumed to have mean 22.5 given
that they come from a normal distribution. To this end, you can use the
function t.test. (You might not know the theory of the t test yet. The
example is mainly included here to give some indication of what "real"
statistical output looks like. A thorough description of t.test is given in
Chapter 4.)

```
> t.test(bmi, mu=22.5)
         One Sample t-test
data:   bmi
t = 0.3449, df = 5, p-value = 0.7442
alternative hypothesis: true mean is not equal to 22.5
95 percent confidence interval:
 18.41734 27.84791
sample estimates:
mean of x
 23.13262
```

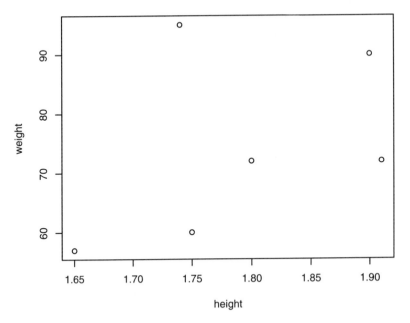

Figure 1.2. A simple x–y plot.

The argument `mu=22.5` attaches a value to the formal argument `mu`, which represents the Greek letter μ conventionally used for the theoretical mean. If this is not given, `t.test` would use the default `mu=0`, which is not of interest here.

For a test like this, we get a more extensive printout than in the earlier examples. The details of the output are explained in Chapter 4, but you might focus on the p-value which is used for testing the hypothesis that the mean is 22.5. The p-value is not small, indicating that it is not at all unlikely to get data like those observed if the mean were in fact 22.5. (Loosely speaking; actually p is the probability of obtaining a t value bigger than 0.3449 or less than -0.3449.) However, you might also look at the 95% confidence interval for the true mean. This interval is quite wide, indicating that we really have very little information about the true mean.

1.1.5 Graphics

One of the most important aspects of presentation and analysis of data is the generation of proper graphics. R has — as S before it — a model for constructing plots that allows simple production of standard plots as well as fine control over the graphical components.

If you want to investigate the relation between weight and height, the first idea is to plot one versus the other. This is done as follows:

```
> plot(height,weight)
```

— leading to Figure 1.2.

You will often want to modify the drawing in various ways. To that end, there is a wealth of plotting parameters that you can set. As an example, let's try changing the plotting symbol, using the keyword pch ("plotting character"), like this:

```
> plot(height, weight, pch=2)
```

This gives the plot in Figure 1.3, with the points now marked with little triangles.

The idea behind the BMI calculation is that this value should be independent of the person's height, thus giving you a single number as an indication of whether someone is overweight and by how much. Since

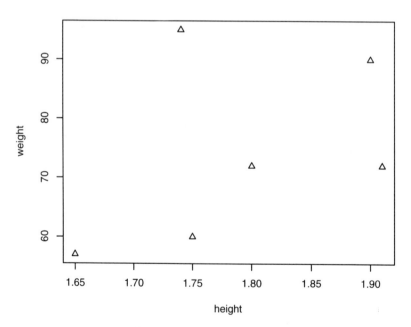

Figure 1.3. Plot with pch = 2.

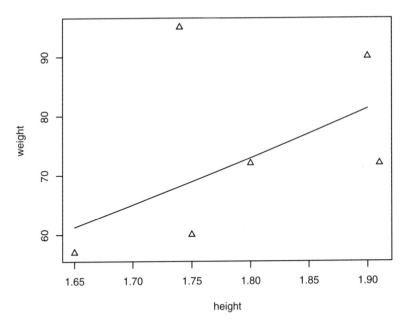

Figure 1.4. Superimposed reference curve, using lines(...).

a normal BMI should be about 22.5, you would expect that *weight* $\approx$ 22.5 $\times$ *height*^{2}. Accordingly, you can superimpose a curve of expected weights at BMI 22.5 on the figure:

```
> hh <- c(1.65, 1.70, 1.75, 1.80, 1.85, 1.90)
> lines(hh, 22.5 * hh^2)
```

yielding Figure 1.4. The function lines will *add* (x, y) values joined by straight lines to an existing plot.

The reason for defining a new variable (hh) with heights rather than using the original height vector is twofold. First, the relation between height and weight is a quadratic one and hence nonlinear although it can be difficult to see on the plot. Since we are approximating a nonlinear curve with a piecewise linear one, it will be better to use points that are spread evenly along the *x*-axis than to rely on the distribution of the original data. Second, since the values of height are not sorted, the line segments would not connect neighbouring points but would run back and forth between distant points.

1.2   R language essentials

This section outlines the basic aspects of the R language. It is necessary to do this in a slightly superficial manner with some of the finer points glossed over. The emphasis is on items that are useful to know in interactive usage as opposed to actual programming, although a brief section on programming is included.

1.2.1   Expressions and objects

The basic interaction mode in R is one of expression evaluation. The user enters an expression; the system evaluates it and prints the result. Some expressions are evaluated not for their result, but for *side effects* such as putting up a graphics window or writing to a file. All R expressions return a value (possibly NULL), but sometimes it is "invisible" and not printed.

Expressions typically involve variable references, operators like +, function calls, as well as some other items that have not been introduced yet.

Expressions work on *objects*. This is an abstract term for anything that can be assigned to a variable. R contains several different types of objects. So far, we have almost exclusively seen numerical vectors, but several other types are introduced in this chapter.

Although objects can be discussed abstractly, it would make a rather boring read without some indication of how to generate them and what to do with them. Conversely, much of the expression syntax makes little sense without knowledge of the objects on which it is intended to work. Therefore, the subsequent sections alternate between introducing new objects and introducing new language elements.

1.2.2   Functions and arguments

At this point you have obtained an impression of the way R works, and we have already used some of the special terminology when talking about the plot *function*, etc. That is exactly the point: Many things in R are done using *function calls*, that is, commands that look like application of a mathematical function of one or several variables, for example, `log(x)` or `plot(height, weight)`.

The format is that a function name is followed by a set of parentheses containing one or more arguments. For instance, in `plot(height,weight)` the function name is `plot` and the arguments are `height` and `weight`.

These are the *actual arguments*, which apply only to the current call. A function also has *formal arguments*, which get connected to actual arguments in the call.

When you write plot (height, weight) R assumes that the first argument corresponds to the *x*-variable and the second one to the *y*-variable. This is known as *positional matching*. Users of traditional programming languages like C or Pascal will recognize this format immediately and probably also be aware of the fact that it becomes unwieldy if a function has a large number of arguments, since you have to supply every one of them and remember their position in the sequence. Fortunately, R has methods to avoid this: Most arguments have sensible defaults and can be omitted in the standard cases, and there are nonpositional ways of specifying them when you need to depart from the default settings.

The plot function is in fact an example of a function that has a large selection of arguments in order to be able to modify symbols, line widths, titles, axis type, and so forth. We used the alternative form of specifying arguments when setting the plot symbol to triangles with plot (height, weight, pch=2).

The pch=2 form is known as a *named actual argument* whose name can be matched against the formal arguments of the function and thereby allow *keyword matching* of arguments. The keyword pch was used to say that the argument is a specification of the plotting character. This type of function argument can be specified in arbitrary order. Thus, you can write plot (y=weight,x=height) and get the same plot as with plot (x=height,y=weight).

The two kinds of argument specification — positional and named — can be mixed in the same call.

Even if there are no arguments to a function call, you have to write, for example, ls () for displaying the contents of the workspace. A common error is to leave off the parentheses, which instead results in the display of a piece of R code since ls entered by itself indicates that you want to see the definition of the function rather than executing it.

The *formal arguments* of a function are part of the function definition. The set of formal arguments to a function, for instance plot.default (which is the function that gets called when you pass plot an x argument for which no special plot method exists), may be seen with

```
> args(plot.default)
function (x, y=NULL, type="p", xlim=NULL, ylim=NULL,
    log="", main=NULL, sub=NULL, xlab=NULL, ylab=NULL,
    ann=par("ann"), axes=TRUE, frame.plot=axes, panel.first=NULL,
    panel.last=NULL, col=par("col"), bg=NA, pch=par("pch"),
    cex=1, lty=par("lty"), lab=par("lab"), lwd=par("lwd"),
```

```
    asp=NA, ...)
NULL
```

Notice that most of the arguments have defaults, meaning that if you do not specify (say) the type argument, the function will behave as if you had passed type="p". The NULL defaults for many of the arguments really serve as indicators that the argument is unspecified, allowing special behaviour to be defined inside the function. For instance, if they are not specified, the xlab and ylab arguments are constructed from the actual arguments passed as x and y. (There are some very fine points associated with this, but we do not go further into the topic.)

The triple-dot (...) argument indicates that this function will accept additional arguments of unspecified name and number. These are often intended for passing on to other functions, although some functions treat it specially. For instance, data.frame interprets the ...-arguments as the column vectors and the argument names become column names in the result.

1.2.3 Vectors

We have already seen numerical vectors. There are two further types, character vectors and logical vectors.

A *character vector* is a vector of text strings, whose elements are specified and printed in quotes:

```
> c("Huey","Dewey","Louie")
[1] "Huey"   "Dewey"  "Louie"
```

It does not matter whether you use single- or double-quote symbols, as long as the left quote is the same as the right quote:

```
> c('Huey','Dewey','Louie')
[1] "Huey"   "Dewey"  "Louie"
```

However, you should avoid the *acute accent* key ('), which is present on some keyboards. Double quotes are used throughout this book to prevent mistakes.

Logical vectors can take the value TRUE or FALSE (or NA; see below). In input, you may use the convenient abbreviations T and F. Logical vectors are constructed using the c function just like the other vector types:

```
> c(T,T,F,T)
[1]  TRUE  TRUE FALSE  TRUE
```

Actually, you will not often have to specify logical vectors in the above manner. It is much more common to use single logical values to turn an option on or off in a function call. Vectors of more than one value most often result from *relational expressions*:

```
> bmi > 25
[1] FALSE FALSE FALSE FALSE   TRUE FALSE
```

We return to relational expressions and logical operations in the context of conditional selection in Section 1.2.11.

1.2.4 Missing values

In practical data analysis a data point is frequently unavailable (the patient did not show up, an experiment failed, etc.). Statistical software needs ways to deal with this. R allows vectors to contain a special NA value. This value is carried through in computations so that operations on NA yield NA as the result. There are some special issues associated with the handling of missing values; we deal with them as we encounter them.

1.2.5 Functions that create vectors

Here we introduce three functions, c, seq, and rep, which are used to create vectors in various situations.

The first of these, c, has already been introduced. It is short for "concatenate", that is, joining items end to end, which is exactly what the function does:

```
> c(42,57,12,39,1,3,4)
[1] 42 57 12 39  1  3  4
```

The second function, seq ("sequence"), is used for equidistant series of numbers. Writing

```
> seq(4,9)
[1] 4 5 6 7 8 9
```

yields, as seen, the integers from 4 to 9. If you want a sequence in jumps of 2, write

```
> seq(4,10,2)
[1]  4  6  8 10
```

This kind of vector is frequently needed, particularly for graphics. For example, we previously used c(1.65,1.70,1.75,1.80,1.85,1.90) to define the *x*-coordinates for a curve, something that could also have been written seq(1.65,1.90,0.05) (the advantage of using seq might have been more obvious if the heights had been in steps of 1 cm rather than 5 cm!).

The case with step size equal to 1 can also be written using a special syntax:

```
> 4:9
[1] 4 5 6 7 8 9
```

The above is exactly the same as seq(4,9), only easier to read.

The third function, rep ("replicate"), is used to generate repeated values. It is used in two variants, depending on whether the second argument is a vector or a single number:

```
> oops <- c(7,9,13)
> rep(oops,3)
[1]  7  9 13  7  9 13  7  9 13
> rep(oops,1:3)
[1]  7  9  9 13 13 13
```

The first of the above function calls repeats the entire vector oops three times. The second call has the number 3 replaced by a vector with the three values (1, 2, 3); these values the values of the oops vector, indicating that 7 should be repeated once, 9 twice, and 13 three times. The rep function is often used for things like group codes: If it is known that the first 10 observations are men and the last 15 ones are women, you can use

```
> rep(1:2,c(10,15))
 [1] 1 1 1 1 1 1 1 1 1 1 2 2 2 2 2 2 2 2 2 2 2 2 2 2 2
```

to form a vector that for each observation indicates whether it is from a man or a woman.

1.2.6 *Matrices and arrays*

A *matrix* in mathematics is just a two-dimensional array of numbers. Matrices are used for many purposes in theoretical and practical statistics, but it is not assumed that the reader is familiar with matrix algebra, so many special operations on matrices, including matrix multiplication, are skipped. (The document "An Introduction to R" that comes with the installation outlines these items quite well.) However, matrices and also higher-dimensional arrays do get used for simpler purposes as well, mainly to hold tables, so an elementary description is in order.

In R, the matrix notion is extended to elements of any type, so you could have, for instance, a matrix of character strings. Matrices and arrays are represented as vectors with dimensions:

```
> x <- 1:12
> dim(x) <- c(3,4)
> x
     [,1] [,2] [,3] [,4]
[1,]    1    4    7   10
[2,]    2    5    8   11
[3,]    3    6    9   12
```

The dim assignment function sets or changes the *dimension attribute* of x, causing R to treat the vector of 12 numbers as a 3 × 4 matrix. Notice that the storage is column-major, that is, the elements of the first column are followed by those of the second, etc.

A convenient way to create matrices is to use the matrix function:

```
> matrix(1:12,nrow=3,byrow=T)
     [,1] [,2] [,3] [,4]
[1,]    1    2    3    4
[2,]    5    6    7    8
[3,]    9   10   11   12
```

Notice how the byrow=T switch causes the matrix to be filled in a rowwise fashion rather than columnwise.

Useful functions that operate on matrices include rownames, colnames and the transposition function t (notice small letter as opposed to capital T for TRUE), which turns rows into columns, and vice versa:

```
> x <- matrix(1:12,nrow=3,byrow=T)
> rownames(x) <- LETTERS[1:3]
> x
  [,1] [,2] [,3] [,4]
A    1    2    3    4
B    5    6    7    8
C    9   10   11   12
> t(x)
     A B  C
[1,] 1 5  9
[2,] 2 6 10
[3,] 3 7 11
[4,] 4 8 12
```

The character vector LETTERS is a built-in variable that contains the capital letters A–Z. Similar useful vectors are letters, month.name, and month.abb with lowercase letters, month names, and abbreviated month names.

You can "glue" vectors together, columnwise or rowwise, using the cbind and rbind functions.

```
> cbind(A=1:4,B=5:8,C=9:12)
     A B  C
[1,] 1 5  9
[2,] 2 6 10
[3,] 3 7 11
[4,] 4 8 12
> rbind(A=1:4,B=5:8,C=9:12)
  [,1] [,2] [,3] [,4]
A    1    2    3    4
B    5    6    7    8
C    9   10   11   12
```

We return to table operations in Section 3.5, which discusses tabulation of variables in a data set.

1.2.7 Factors

It is common in statistical data to have categorical variables, indicating some subdivision of data, such as social class, primary diagnosis, tumor stage, Tanner stage of puberty, etc. Typically, these are input using a numeric code.

Such variables should be specified as *factors* in R. This is a data structure that (among other things) makes it possible to assign meaningful names to the categories.

There are analyses where it is essential for R to be able to distinguish between categorical codes and variables whose values have a direct numerical meaning (see Chapter 6).

The terminology is that a factor has a set of *levels* — say four levels for concreteness. Internally, a four-level factor consists of two items: (a) a vector of integers between 1 and 4 and (b) a character vector of length 4 containing strings describing what the four levels are. Let's look at an example:

```
> pain <- c(0,3,2,2,1)
> fpain <- factor(pain,levels=0:3)
> levels(fpain) <- c("none","mild","medium","severe")
```

The first command creates a numerical vector pain, encoding the pain level of five patients. We wish to treat this as a categorical variable, so we create a factor fpain from it using the function factor. This is called with one argument in addition to pain, namely levels=0:3, which indicates that the *input* coding uses the values 0–3. The latter can in principle be left out, since R by default uses the values in pain, suitably sorted, but

it is a good habit to retain it; cf. below. The effect of the final line is that the
level names are changed to the four specified character strings.

The result should be apparent from the following:

```
> fpain
[1] none    severe medium medium mild
Levels:   none mild medium severe
> as.numeric(fpain)
[1] 1 4 3 3 2
> levels(fpain)
[1] "none"    "mild"    "medium" "severe"
```

The function `as.numeric` extracts the numerical coding as numbers
1–4 and `levels` extracts the names of the levels. Notice that the original input coding in terms of numbers 0–3 has disappeared; the internal
representation of a factor always uses numbers starting at 1.

If, in `factor(...)`, you do not specify a `levels` argument, the levels
will by default be the sorted, unique values represented in the vector. This
is not always desirable when dealing with text variables, since the sorting
is *alphabetical*. Consider, for instance,

```
> text.pain <-  c("none","severe",  "medium",  "medium",  "mild")
> factor(text.pain)
[1] none    severe medium medium mild
Levels:   medium mild none severe
```

Also, you would obviously not include levels that are not present in data,
which can cause trouble when merging data from several populations.

R also allows you to create a special kind of factor in which the levels are ordered. This is done using the `ordered` function, which works
similarly to `factor`. These are potentially useful in that they distinguish
nominal and ordinal variables from each other (and arguably, `text.pain`
above ought to have been an ordered factor). Unfortunately, R defaults to
treating the levels as if they were *equidistant* in the modelling code (by generating polynomial contrasts), so it may be better to ignore ordered factors
for now.

1.2.8 Lists

It is sometimes useful to combine a collection of objects into a larger
composite object. This can be done using *lists*.

You can construct a list from its components with the function `list`.

As an example, consider a set of data from Altman (1991, p. 183) concerning pre- and postmenstrual energy intake in a group of women. We can place these data in two vectors as follows:

```
> intake.pre <- c(5260,5470,5640,6180,6390,
+ 6515,6805,7515,7515,8230,8770)
> intake.post <- c(3910,4220,3885,5160,5645,
+ 4680,5265,5975,6790,6900,7335)
```

Notice how input lines can be broken and continue on the next line. If you press the Enter key while an expression is syntactically incomplete, R will assume that the expression continues on the next line and will change its normal > prompt to the *continuation prompt* +. This often happens inadvertently due to a forgotten parenthesis or a similar problem; in such cases either complete the expression on the next line, or press ESC (Windows) or Ctrl-C (Unix).

To combine these individual vectors into a list, you can say

```
> mylist <- list(before=intake.pre,after=intake.post)
> mylist
$before
 [1] 5260 5470 5640 6180 6390 6515 6805 7515 7515 8230 8770

$after
 [1] 3910 4220 3885 5160 5645 4680 5265 5975 6790 6900 7335
```

The components of the list are named according to the argument names used in `list`. Named components may be extracted like this:

```
> mylist$before
 [1] 5260 5470 5640 6180 6390 6515 6805 7515 7515 8230 8770
```

Many of R's built-in functions compute more than a single vector of values and return their results in the form of a list.

1.2.9   Data frames

A data frame corresponds to what other statistical packages call a "data matrix" or a "data set". It is a list of vectors and/or factors of the same length, which are related "across", such that data in the same position come from the same experimental unit (subject, animal, etc.). In addition, it has a unique set of row names.

You can create data frames from preexisting variables:

```
> d <- data.frame(intake.pre,intake.post)
> d
```

```
   intake.pre intake.post
1         5260        3910
2         5470        4220
3         5640        3885
4         6180        5160
5         6390        5645
6         6515        4680
7         6805        5265
8         7515        5975
9         7515        6790
10        8230        6900
11        8770        7335
```

Notice that these data are paired, that is, the same woman has an intake of 5260 kJ premenstrually and 3910 kJ postmenstrually.

As with lists, variables are accessible using the $ notation:

```
> d$intake.pre
 [1] 5260 5470 5640 6180 6390 6515 6805 7515 7515 8230 8770
```

1.2.10   Indexing

If you need a particular element in a vector, for instance the premenstrual energy intake for woman no. 5, you can do

```
> intake.pre[5]
[1] 6390
```

The brackets are used for selection of data, also known as *indexing* or *subsetting*. This also works on the left-hand side of an assignment (so that you can say, for instance, intake.pre[5] <- 6390) if you want to modify elements of a vector.

If you want a subvector consisting of data for more than one woman, let's say nos. 3, 5, and 7, you can index with a vector:

```
> intake.pre[c(3,5,7)]
[1] 5640 6390 6805
```

Note that it is necessary to use the c(...)-construction to define the vector consisting of the three numbers 3, 5, and 7. intake.pre[3,5,7] would mean something completely different. It would specify indexing into a three-dimensional array.

Of course, indexing with a vector also works if the index vector is stored in a variable. This is useful when you need to index several variables in the same way.

```
> v <- c(3,5,7)
> intake.pre[v]
[1] 5640 6390 6805
```

It is also worth noting that to get a sequence of elements, for instance, the first five, you can use the a : b notation:

```
> intake.pre[1:5]
[1] 5260 5470 5640 6180 6390
```

A neat feature of R is the possibility of negative indexing. You can get all observations *except* nos. 3, 5, and 7 by writing

```
> intake.pre[-c(3,5,7)]
[1] 5260 5470 6180 6515 7515 7515 8230 8770
```

It is not possible to mix positive and negative indices. That would be highly ambiguous.

1.2.11 Conditional selection

We saw in Section 1.2.10 how to extract data using one or several indices. In practice, you often need to extract data that satisfy certain criteria, such as data from the males or the prepubertal or those with chronic diseases, etc. This can be done simply by inserting a relational expression instead of the index, like this:

```
> intake.post[intake.pre > 7000]
[1] 5975 6790 6900 7335
```

— yielding the postmenstrual energy intake for the four women who had an energy intake above 7000 kJ premenstrually.

Of course, this kind of expression makes sense only if the variables that go into the relational expression have the same length as the variable being indexed.

The comparison operators available are < (less than), > (greater than), == (equal to), <= (less than or equal to), >= (greater than or equal to), and != (not equal to). Notice that a double equal sign is used for testing equality. This is to avoid confusion with the = symbol used to match keywords with function arguments. Also, the != operator is new to some; the ! symbol indicates negation. The same operators are used in the C programming language.

To combine several expressions you can use the logical operators & (logical "and"), | (logical "or"), and ! (logical "not"). For instance, we find the postmenstrual intake for women with a premenstrual intake between 7000 and 8000 kJ with

```
> intake.post[intake.pre > 7000 & intake.pre <= 8000]
[1] 5975 6790
```

There are also `&&` and `||`, which are used for flow control in actual R programming. However, their use is beyond what we discuss here.

It may be worth taking a closer look at what actually happens when you use a logical expression as an index. The result of the logical expression is a logical vector as described in Section 1.2.3:

```
> intake.pre > 7000 & intake.pre <= 8000
 [1] FALSE FALSE FALSE FALSE FALSE FALSE FALSE  TRUE  TRUE  FALSE
[11] FALSE
```

Indexing with a logical vector implies that you pick out the values where the logical vector is TRUE, so in the preceding example we got the 8th and 9th values in `intake.post`.

If missing values (NA; see Section 1.2.4) appear in an indexing vector, then R will create the corresponding elements in the result but set the values to NA.

In addition to the relational and logical operators, there is a series of functions that return a logical value. A particularly important one is `is.na(x)`, which is used to find out which elements of x are recorded as missing (NA).

Notice that there is a real need for `is.na` because you cannot make comparisons of the form x==NA. That simply gives NA as the result, for any value of x. The result of a comparison with an unknown value is unknown!

1.2.12   *Indexing of data frames*

We have already seen how it is possible to extract variables from a data frame by typing, for example, `d$intake.post`. However, it is also possible to use a notation that uses the matrix-like structure directly:

```
> d <- data.frame(intake.pre,intake.post)
> d[5,1]
[1] 6390
```

gives fifth row, first column, that is, the "pre" measurement for woman no. 5, and

```
> d[5,]
  intake.pre intake.post
5       6390        5645
```

gives *all* measurements for woman no. 5. Notice that the comma in d[5,] is required; without the comma, for example d[2], you get the data frame consisting of the second *column* of d, that is, more like d[,2], which is the column itself.

Other indexing techniques also apply. In particular, it can be useful to extract all data for cases that satisfy some criterion, such as women with a premenstrual intake above 7000 kJ:

```
> d[d$intake.pre>7000,]
   intake.pre intake.post
8        7515        5975
9        7515        6790
10       8230        6900
11       8770        7335
```

Here we extracted the rows of the data frame where intake.pre>7000. Notice that the row names are those of the original data frame.

If you want to understand the details of this, it may be a little easier if it is divided into smaller steps. It could also have been done like this:

```
> sel <- d$intake.pre>7000
> sel
 [1] FALSE FALSE FALSE FALSE FALSE FALSE FALSE   TRUE   TRUE   TRUE
[11]   TRUE
> d[sel,]
   intake.pre intake.post
8        7515        5975
9        7515        6790
10       8230        6900
11       8770        7335
```

What happens is that sel *(select)* becomes a logical vector that is TRUE corresponding to the four women consuming more than 7000 kJ premenstrually. Indexing as d[sel,] yields data from the rows where sel is TRUE and from all columns because of the empty field after the comma.

1.2.13 subset and transform

The indexing techniques for extracting parts of a data frame are logical but a bit cumbersome, and a similar comment applies to the process of adding transformed variables to a data frame. Therefore, a couple of functions

exist to make things a little easier. They are used as follows (data is used
to fetch a built-in data set; see Section 1.5.4):

```
> data(thuesen)
> thue2 <- subset(thuesen,blood.glucose<7)
> thue2
   blood.glucose short.velocity
6            5.3           1.49
11           6.7           1.25
12           5.2           1.19
15           6.7           1.52
17           4.2           1.12
22           4.9           1.03
> thue3 <- transform(thuesen,log.gluc=log(blood.glucose))
> thue3
   blood.glucose short.velocity log.gluc
1           15.3           1.76 2.727853
2           10.8           1.34 2.379546
3            8.1           1.27 2.091864
4           19.5           1.47 2.970414
5            7.2           1.27 1.974081
...
22           4.9           1.03 1.589235
23           8.8           1.12 2.174752
24           9.5           1.70 2.251292
```

Notice that the variables used in the expressions for new variables or for
subsetting are evaluated with variables taken from the data frame.

subset also works on single vectors. This is nearly the same as indexing
with a logical vector (such as short.velocity[blood.glucose<7]),
except that observations with missing values in the selection criterion are
excluded.

subset also has a select formal argument with a somewhat peculiar
syntax, used to extract variables from a data frame. This will not be needed
here, though.

1.2.14 Grouped data and data frames

The natural way of storing grouped data in a data frame is to have the
data themselves in one vector and parallel to that to have a factor telling
which data are from which group. Consider, for instance, the following
data set on energy expenditure for lean and obese women. (Again, see
Section 1.5.4 for data.)

```
> data(energy)
> energy
   expend stature
```

```
1      9.21    obese
2      7.53     lean
3      7.48     lean
4      8.08     lean
5      8.09     lean
6     10.15     lean
7      8.40     lean
8     10.88     lean
9      6.13     lean
10     7.90     lean
11    11.51    obese
12    12.79    obese
13     7.05     lean
14    11.85    obese
15     9.97    obese
16     7.48     lean
17     8.79    obese
18     9.69    obese
19     9.68    obese
20     7.58     lean
21     9.19    obese
22     8.11     lean
```

This is a convenient format since it generalizes easily to multiple classification criteria. However, sometimes it is desirable to have data in a separate vector for each group. Fortunately, it is easy to extract these from the data frame:

```
> exp.lean <- energy$expend[energy$stature=="lean"]
> exp.obese <- energy$expend[energy$stature=="obese"]
```

Alternatively, you can use the split function, which generates a list of vectors according to a grouping.

```
> l <- split(energy$expend, energy$stature)
> l
$lean
 [1]  7.53  7.48  8.08  8.09 10.15  8.40 10.88  6.13  7.90  7.05
[11]  7.48  7.58  8.11

$obese
[1]  9.21 11.51 12.79 11.85  9.97  8.79  9.69  9.68  9.19
```

1.2.15 Sorting

It is trivial to sort a vector. Just use the sort function:

```
> intake.post
 [1] 3910 4220 3885 5160 5645 4680 5265 5975 6790 6900 7335
> sort(intake.post)
```

```
[1] 3885 3910 4220 4680 5160 5265 5645 5975 6790 6900 7335
```

(intake.pre could not be used for this example since it is sorted already!)

However, sorting a single vector is not usually what is required. More often, you need to sort a series of variables according to the values of some *other* variables — blood pressures sorted by sex and age, for instance. For this purpose, there is a construction that may look somewhat abstract at first but is really very powerful. You first compute an *ordering* of a variable.

```
> order(intake.post)
 [1]  3  1  2  6  4  7  5  8  9 10 11
```

The result is the numbers 1 to 11 (or whatever the length of the vector is), sorted according to the size of the argument to order (here: intake.post). Interpreting the result of order is a bit tricky — it should be read as follows: You sort intake.post by placing its values in the order no. 3, no. 1, no. 2, no. 6, etc.

The point is that by indexing with this vector, other variables can be sorted by the same criterion. Note that indexing with a vector containing the numbers from 1 to the number of elements exactly once corresponds to a reordering of the elements.

```
> o <- order(intake.post)
> intake.post[o]
 [1] 3885 3910 4220 4680 5160 5265 5645 5975 6790 6900 7335
> intake.pre[o]
 [1] 5640 5260 5470 6515 6180 6805 6390 7515 7515 8230 8770
```

What has happened here is that intake.post has been sorted — just as in sort(intake.post) — while intake.pre has been sorted by the size of the corresponding intake.post.

Sorting by several criteria is done simply by having several arguments to order; for instance, order(sex,age) will give a main division into men and women, and within each sex an ordering by age. The second variable is used when the order cannot be decided from the first variable. Sorting in reverse order can be handled by, for example, changing the sign of the key variable.

1.2.16 Implicit loops

The looping constructs of R are described in Section 1.4.1. For the purposes
of this book, you can largely ignore their existence. However, there is a
group of R functions that it will be useful for you to know about.

A common application of loops is to apply a function to each element of a
set of values or vectors and collect the results in a single structure. In R this
is abstracted by the functions lapply and sapply. The former always
returns a list (hence the 'l') whereas the latter tries to simplify (hence the
's') the result to a vector or a matrix if possible. So to compute the mean of
each variable in a data frame of numeric vectors, you can do the following:

```
> lapply(thuesen, mean, na.rm=T)
$blood.glucose
[1] 10.3

$short.velocity
[1] 1.325652

> sapply(thuesen, mean, na.rm=T)
 blood.glucose short.velocity
      10.300000       1.325652
```

Notice how both forms attach meaningful names to the result, which is another good reason to prefer using these functions over using explicit loops.
The second argument to lapply/sapply is the function that should be
applied, here mean. Any further arguments are passed on to the function;
in this case we pass na.rm=T to request that missing values be removed
(see Section 3.1).

A similar function, apply, allows you to apply a function to the rows
or columns of a matrix (or over indices of a multidimensional array in
general) as in

```
> m <- matrix(rnorm(12),4)
> m
             [,1]        [,2]       [,3]
[1,] -0.9686756 -0.7045067 0.9612115
[2,]  0.9658574  0.2166490 0.3189893
[3,]  1.9969486  1.4072639 0.4043070
[4,] -1.1616653 -0.7892404 2.0291297
> apply(m, 2, min)
[1] -1.1616653 -0.7892404  0.3189893
```

The second argument is the index (or vector of indices) that defines what
the function is applied to; in this case we get the columnwise minima.

Also, the function tapply allows you to create tables (hence the 't') of the
value of a function on subgroups defined by its second argument, which

can be a factor or a list of factors. In the latter case a cross-classified table is generated. (The grouping can also be defined by ordinary vectors. They will be converted to factors internally.)

```
> tapply(energy$expend, energy$stature, median)
 lean obese
 7.90  9.69
```

1.3   The graphics subsystem

In Section 1.1.5 we saw how to generate a simple plot and superimpose a curve on it. It is quite common in statistical graphics that you want to create a plot that is slightly different from the default: Sometimes you will want to add annotation, sometimes you want the axes to be different — labels instead of numbers, irregular placement of tick marks, etc. All these things can be obtained in R. The methods for doing it may feel slightly unusual at first, but it is a very flexible and powerful approach.

In this section we look deeper into the structure of a typical plot and give some indication of how you can work with plots to achieve your desired results. Beware, though, that this is a large and complex area and it is not within the scope of this book to cover it completely.

1.3.1   Plot layout

In the graphics model that R uses, there is (for a single plot) a figure region containing a central plotting region surrounded by margins. Coordinates inside the plotting region are specified in data units (the kind generally used to label the axes). Coordinates in the margins are specified in *lines of text* as you move in a direction perpendicular to a side of the plotting region, but in data units as you move along the side. This is useful since you generally want to put text in the margins of a plot.

A standard *x–y* plot has an *x* and a *y* title label generated from the expressions being plotted. You may, however, override these labels and also add two further titles, a main title above the plot and a subtitle at the very bottom, in the plot call.

```
> x <- runif(50,0,2)
> y <- runif(50,0,2)
> plot(x, y, main="Main title", sub="subtitle",
+      xlab="x-label", ylab="y-label")
```

Inside the plotting region you can place points and lines that are either specified in the `plot` call or added later with `points` and `lines`. You can also place a text with

```
> text(0.6,0.6,"text at (0.6,0.6)")
> abline(h=.6,v=.6)
```

Here, the `abline` call is just to show how the text is centered on the point $(0.6, 0.6)$. (Normally, `abline` plots the line $y = a + bx$ when given a and b as arguments, but it can also be used to draw horizontal and vertical lines as shown.)

The margin coordinates are used by the `mtext` function. They can be demonstrated as follows:

```
> for (side in 1:4) mtext(-1:4,side=side,at=.7,line=-1:4)
> mtext(paste("side",1:4),  side=1:4,  line=-1,font=2)
```

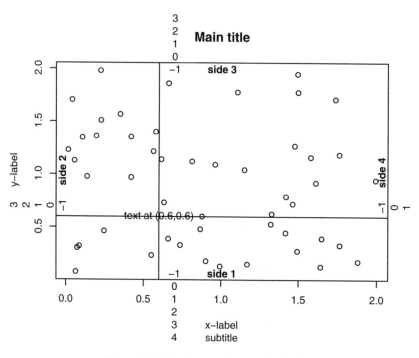

Figure 1.5. The layout of a standard plot.

The `for` loop (see Section 1.4.1) places the numbers -1 to 4 on corresponding lines in each of the four margins, at an off-center position of 0.7 measured in user coordinates. The subsequent call places a label on each side, giving the side number. The argument `font=2` means that a boldface font is used. Notice in Figure 1.5 that the margins are not all wide enough to hold all the numbers and that it is possible to use negative line numbers to place text within the plotting region.

1.3.2   Building a plot from pieces

High-level plots are composed of elements, each of which can also be drawn separately. The separate drawing commands often allow finer control of the element, so a standard strategy to achieve a given effect is first to draw the plot without that element and add the element subsequently. As an extreme case, the following command will plot absolutely nothing:

```
> plot(x, y, type="n", xlab="", ylab="", axes=F)
```

Here type="n" causes the points not to be drawn. axes=F suppresses the axes and the box around the plot, and the x and y title labels are set to empty strings.

However, the fact that nothing is plotted does not mean that nothing happened. The command sets up the plotting region and coordinate systems just as if it had actually plotted the data. To add the plot elements, evaluate the following:

```
> points(x,y)
> axis(1)
> axis(2,at=seq(0.2,1.8,0.2))
> box()
> title(main="Main title", sub="subtitle",
+      xlab="x-label", ylab="y-label")
```

Notice how the second axis call specifies an alternative set of tick marks (and labels). This is a common technique used to create special axes on a plot and might also be used to create nonequidistant axes as well as axes with nonnumeric labelling.

Plotting with type="n" is a common technique in S-PLUS, which lacks R's feature of passing a vector argument for col to specify individual colours for each point. Instead, to create a plot with different colours for different groups, you would first plot all data with type="n" to make sure the plot region was large enough, and then you would add the points for each group using points.

1.3.3 Using par

The par function allows incredibly fine control over the details of a plot, although it can be quite confusing to the beginner (and even to experienced users at times). The best strategy for learning it may well be simply to try and pick up a few useful tricks at a time and once in a while try to solve a particular problem by poring over the help page.

Some of the parameters, but not all, can also be set via arguments to plotting functions, which also have some arguments that cannot be set by par. When a parameter can be set by both methods, the difference is generally that if something is set via par, then it stays set subsequently.

The par settings allow you to control line widths and types, character size and font, colours, style of axis calculation, size of the plot and figure regions, clipping, etc. It is possible to divide a figure into several subfigures, using the mfrow and mfcol parameters.

For instance, the default margin sizes are just over 5, 4, 4, and 2 lines. You might set par(mar=c(4,4,2,2)+0.1) before plotting. This shaves one line off the bottom margin and two lines off the top margin of the plot, which will reduce the amount of unused whitespace when there is no main title or subtitle. If you look carefully, you will in fact notice that Figure 1.5 has a somewhat smaller plotting region than the other plots in this book. This is because the other plots have been made with reduced margins for typographical reasons.

However, it is quite pointless to describe the graphics parameters completely at this point. Instead, we return to them as they are used for specific plots.

1.3.4 Combining plots

Some special considerations arise when you wish to put several elements together in the same plot. Consider overlaying a histogram with a normal density (see Sections 3.2 and 3.4.1 for information on histograms and Section 2.5.1 for density). The following is close, but only nearly good enough (figure not shown).

```
> x <- rnorm(100)
> hist(x,freq=F)
> curve(dnorm(x),add=T)
```

The freq=F argument to hist ensures that the histogram is in terms of densities rather than absolute counts. The curve function graphs an expression (in terms of x) and its add=T allows it to overplot an existing plot. So things are generally set up correctly, but sometimes the top of the density function gets chopped off. The reason is of course that the height of the normal density played no role in the setting of the y-axis for the histogram. It will not help to reverse the order and draw the curve first and add the histogram, because then the highest bars might get clipped.

The solution is first to get hold of the magnitude of the y values for both plot elements and make the plot big enough to hold both (Figure 1.6):

```
> h <- hist(x, plot=F)
> ylim <- range(0, h$density, dnorm(0))
> hist(x, freq=F, ylim=ylim)
> curve(dnorm(x), add=T)
```

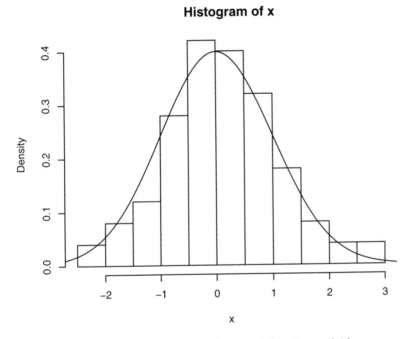

Figure 1.6. Histogram with normal density overlaid.

When called with `plot=F`, `hist` will not plot anything, but it will return a structure containing the bar heights on the density scale. This and the fact that the maximum of `dnorm(x)` is `dnorm(0)` allows us to calculate a range covering both the bars and the normal density. The zero in the `range` call ensures that the bottom of the bars will be in range too. The range of y values is then passed to the `hist` function via the `ylim` argument.

1.4 R programming

It is possible to write your own R functions. In fact, this is a major aspect and attraction of working with the system in the long run. This book largely avoids the issue in favour of covering a larger set of basic statistical procedures that can be executed from the command line. However, to give you a feel for what can be done, consider the following function, which wraps the code from the example of Section 1.3.4 so that you can just say hist.with.normal(rnorm(200)). It has been slightly extended so that it now uses the empirical mean and standard deviation of the data instead of just 0 and 1.

```
> hist.with.normal <- function(x, xlab=deparse(substitute(x)),...)
+ {
+     h <- hist(x, plot=F, ...)
+     s <- sd(x)
+     m <- mean(x)
+     ylim <- range(0,h$density,dnorm(0,sd=s))
+     hist(x, freq=F, ylim=ylim, xlab=xlab, ...)
+     curve(dnorm(x,m,s), add=T)
+ }
```

Notice the use of a default argument for xlab. If xlab is not specified, then it is obtained from this expression which evaluates to a character form of the expression given for x, that is, if you pass rnorm(100) for x, then the x label becomes "rnorm(100)". Notice also the use of a ... argument which collects any additional arguments and passes them on to hist in the two calls.

You can learn more about programming in R by studying the built-in functions, starting with simple ones like log10 or weighted.mean. For further study of the R and S languages, the books by Venables and Ripley (2002, 2000) are indispensable as are the original books on (new) S, known as the "blue book" (Becker et al., 1988) and the "white book" (Chambers and Hastie, 1992).

1.4.1 Flow control

Until now, we have seen components of the R language that cause evaluation of single expressions. However, R is a true programming language that allows conditional execution and looping constructs as well. Consider, for instance, the following code. (It is not terribly important what the code does, but it implements a version of Newton's method for calculating the square root of y.)

```
> y <- 12345
```

```
> x <- y/2
> while (abs(x*x-y) > 1e-10) x <- (x + y/x)/2
> x
[1] 111.1081
> x^2
[1] 12345
```

Notice the `while(condition) expression` construction, which says that the expression should be evaluated as long as the condition is TRUE. The test occurs at the top of the loop so that the expression might never be evaluated.

A variation of the same algorithm with the test at the bottom of the loop can be written with a `repeat` construction:

```
> x <- y/2
> repeat{
+       x <- (x + y/x)/2
+       if (abs(x*x-y) < 1e-10) break
+ }
> x
[1] 111.1081
```

This also illustrates three other flow control structures: (a) a *compound expression*: several expressions held together between curly braces; (b) an `if` construction for conditional execution; and (c) a `break` expression, which causes the enclosing loop to exit.

Incidentally, the loop could allow for y being a vector, simply by changing the termination condition to

```
if (all(abs(x*x - y) < 1e-10)) break
```

This would iterate excessively for some elements, but the vectorized arithmetic would likely more than make up for that.

Actually, `while` and `repeat` are quite rarely used in R. Much more frequent is `for`, which loops over a fixed set of values as in the following example, which plots a set of power curves on the unit interval.

```
> x <- seq(0, 1, .05)
> plot(x, x, ylab="y", type="l")
> for ( j in 2:8 ) lines(x, x^j)
```

Notice the *loop variable* j, which in turn takes the values of the given sequence when used in the `lines` call.

1.4.2   Classes and generic functions

Object-oriented programming is about creating coherent systems of data and methods that work upon them. One purpose is to simplify programs by accommodating the fact that you will have conceptually similar methods for different types of data, even though the implementations will have to be different. A prototype example is the print method: It generally makes sense to print many kinds of data objects, but the print layout will depend on what the data object is. You will generally have a *class* of data objects and a *print method* for that class. There are several object-oriented languages implementing these ideas in different ways.

R uses the same object system as S version 3. This is a simple system in which an object has a `class` attribute, which is simply a character vector. One example of this is that all the return values of the classical tests like `t.test` have class `"htest"`, indicating that they are the result of a hypothesis test. When these objects are printed, it is done by `print.htest`, which creates the nice layout (see Chapter 4 for examples). However, from a programmatic viewpoint these objects are just lists and you can, for instance, extract the *p*-value by writing

```
> t.test(bmi, mu=22.5)$p.value
[1] 0.7442183
```

The function `print` is a *generic function*, one that acts differently depending on its argument. These are generally implemented as follows:

```
> print
function (x, ...)
UseMethod("print")
```

which means that R should pass control to a function named according to the object class (`print.htest` for objects of class `"htest"`, etc.) or if this is not found, to `print.default`. To see all the methods available for `print`, type `methods(print)` (there are 58 of them in 1.5.0, so the output is not shown here).

1.5   Session management

1.5.1   The workspace

All variables created in R are stored in a common workspace. To see which variables are defined in the workspace, you can use the function `ls` (*list*).

It should look as follows if you have run all the preceding examples in this chapter:

```
> ls()
 [1] "bmi"               "d"              "energy"
 [4] "exp.lean"          "exp.obese"      "fpain"
 [7] "h"                 "height"         "hh"
[10] "hist.with.normal"  "intake.post"    "intake.pre"
[13] "j"                 "l"              "m"
[16] "mylist"            "o"              "oops"
[19] "pain"              "pu"             "sel"
[22] "side"              "text.pain"      "thue2"
[25] "thue3"             "thuesen"        "v"
[28] "weight"            "x"              "xbar"
[31] "y"                 "ylim"
```

Remember that you cannot omit the parentheses in `ls()`. An alternative that might be preferable, especially if you plan to use S-PLUS later on, is `objects()`. In R it is just another name for the same function.

If at some point things begin to look messy, you can delete some of the objects. This is done using `rm` (*remove*), so that

```
> rm(height, weight, bmi)
```

deletes the variables `height`, `weight`, and `bmi`.

The entire workspace can be cleared using `rm(list=ls())`, and in the Windows version also via the "Remove all objects" menu entry (under "Misc"). Strictly speaking, you need `rm(list=ls(all.names=TRUE))` since `ls` otherwise does not list variables whose name begins with a dot. However, because such variable names are often used for system purposes, you are discouraged from using such names in the first place.

If you are acquainted with the Unix operating system, for which the S language that preceded R was originally written, then you will know that the commands for listing and removing files in Unix are called precisely `ls` and `rm`.

It is possible to save the workspace to a file at any time. If you just write

```
save.image()
```

then it will be saved to a file called `.RData` in your working directory. The Windows version also has this on the File menu. When you exit R you will be asked whether to save the workspace image; if you accept, the same thing will happen. It is also possible to specify an alternative filename (within quotes). You can also save selected objects with `save`. The `.RData` file is loaded by default when R is started in its directory. Other save files can be loaded into your current workspace using `load`.

It is important to note that the workspace consists only of R objects, not of any of the output that you have generated during a session. If you want to save your output, use "Save to file" from the File menu on Windows or use standard cut-and-paste facilities. You can also use ESS, which works on both Unix/Linux systems and on Windows. It is a "mode" for the Emacs editor where you can run your entire session in an Emacs buffer. You can get ESS and installation instructions for it from CRAN (see Appendix A).

The history of commands entered in a session can be saved and reloaded using the `savehistory` and `loadhistory` commands, which are also mapped to menu entries on Windows.

1.5.2 Getting help

R can do a lot more than what a typical beginner can be expected to need or even understand. This book is written so that most of the code you are likely to need in relation to the statistical procedures is described in the text, and the compendium in Appendix C is designed to provide a basic overview. However, it is obviously not possible to cover everything.

R also comes with extensive online help in text form as well as in the form of a series of HTML files that can be read using a Web browser such as Netscape or Internet Explorer. The help pages can be accessed via "help" in the menu bar on Windows and by entering `help.start()` on any platform. You will find that the pages are of a technical nature. Precision and conciseness here take precedence over readability and pedagogy (something one learns to appreciate after exposure to the opposite).

From the command line, you can always enter `help(aggregate)` to get help on the `aggregate` function, or use the prefix form `?aggregate`. If the HTML viewer is running, then the help page is shown there. Otherwise it is shown as text either through a pager to the terminal window or in a separate window.

Notice that the HTML version of the help system features a very useful "Search Engine and Keywords" and that the `apropos` function allows you to get a list of command names that contain a given pattern.

Also available with the R distributions is a set of documents in various formats. Of particular interest is "An Introduction to R", originally based on a set of notes for S-PLUS by Bill Venables and David Smith and modified for R by various people. It contains an introduction to the R language and environment in a rather more language-centric fashion than this book. On the Windows platform, you can choose to install PDF documents as part of the installation procedure so that — provided that the Adobe Acrobat Reader program is also installed — it can be accessed via the Help menu.

An HTML version (without pictures) can be accessed via the browser interface on all platforms.

1.5.3   Packages

An R installation contains a library of packages. Some of these packages are part of the basic installation. Others can be downloaded from CRAN (see Appendix A), which currently hosts over 100 packages for various purposes. You can even create your own packages.

A package can contain functions written in the R language, dynamically loaded libraries of compiled code (written in C or Fortran mostly), and data sets. It generally implements functionality that most users will probably not need to have loaded all the time. A package is loaded into R using the library command, so to load the survival package you should enter

```
> library(survival)
```

The loaded packages are not considered part of the user workspace. If you terminate your R session and start a new session with the saved workspace, then you will have to load the packages again. For the same reason, it is rarely necessary to remove a package that you have loaded, but it can be done if desired with

```
> detach("package:survival")
```

(see also Section 1.5.5).

1.5.4   Built-in data

We have used the data function without explanation a couple of times already. It is used to load a built-in data set (one that comes with R or one of the packages) into memory. Most often, it loads a data frame with the name that its argument specifies; data(thuesen) will, for instance, load the thuesen data frame. However, it may be another kind of object, or even several objects.

What data does is to go through the data directories associated with each package (see Section 1.5.3) and look for files whose basename matches the given name. Depending on the file extension several things can then happen. Files with a .tab extension are read using read.table (Section 1.6), whereas files with a .R extension are executed as source files (and could, in general, do anything!), to give two common examples. If there is a subdirectory of the current directory called data, then it is searched as well.

1.5.5 *attach and detach*

The notation for accessing variables in data frames gets rather heavy if you repeatedly have to write longish commands like

```
plot(thuesen$blood.glucose,thuesen$short.velocity)
```

Fortunately, you can make R look for objects among the variables in a given data frame, for example thuesen. You write

```
> attach(thuesen)
```

and then thuesen's data are available without the clumsy $-notation:

```
> blood.glucose
 [1] 15.3 10.8  8.1 19.5  7.2  5.3  9.3 11.1  7.5 12.2  6.7  5.2
[13] 19.0 15.1  6.7  8.6  4.2 10.3 12.5 16.1 13.3  4.9  8.8  9.5
```

What happens is that the data frame thuesen is placed in the system's *search path*. You can view the search path with search:

```
> search()
[1] ".GlobalEnv"      "thuesen"       "package:ISwR"
[4] "package:ctest"  "Autoloads"     "package:base"
```

Notice that thuesen is placed as no. 2 in the search path. .GlobalEnv is the workspace and package:base is the system library where all standard functions are defined. Autoloads is not described here. package:ctest contains the "classical tests": the Wilcoxon test, etc. Finally, package:ISwR contains data sets used in this book.

There may be several objects of the same name in different parts of the search path. In that case, R chooses the first one (that is, it searches first in .GlobalEnv, then in thuesen, and so forth). For this reason you need to be a little careful with "loose" objects that are defined in the workspace outside a data frame since they will be used before any vectors and factors of the same name in an attached data frame. For the same reason, it is not a good idea to give a data frame the same name as one of the variables inside it. Note also that changing a data frame after attaching it will not affect the variables available since attach involves a (virtual) copy operation of the data frame.

It is not possible to attach data frames in front of .GlobalEnv or following package:base. However, it is possible to attach more than one data frame. New data frames are inserted into position 2 by default, and everything except .GlobalEnv moves one step to the right. It is, however, possible to specify that a data frame should be searched before .GlobalEnv by using constructions of the form

```
with(thuesen, plot(blood.glucose, short.velocity), thuesen)
```

In some contexts, R uses a slightly different method when looking for objects. If the program "knows" that it needs a variable of a specific type (usually a function), it will skip those of other types. This is what saves you from the worst consequences of accidentally naming a variable (say) c, even though there is a system function of the same name.

You can remove a data frame from the search path with detach. If no arguments are given, the data frame in position 2 is removed, which is generally what is desired. .GlobalEnv and package:base cannot be detach'ed.

```
> detach()
> search()
[1] ".GlobalEnv"     "package:ISwR"     "package:ctest"
[4] "Autoloads"      "package:base"
```

1.6   Data entry

Data sets do not have to be very large before it becomes impractical to type them in with c(...). In this section we discuss how to read data files and how to use the data editor module in R. The text has some bias toward Windows systems, mainly because this is where problems are most frequently encountered.

1.6.1   Reading from a text file

The most convenient way of reading data into R is via the function called read.table. It requires that data is in "ASCII format", that is, a "flat file" as created with Windows' NotePad or any plain-text editor. The result of read.table is a data frame, and it expects to find data in a corresponding layout where each line in the file contains all data from one subject (or rat or ...) in a specific order, separated by blanks or, optionally, some other separator. The first line of the file can contain a header giving the names of the variables, a practice that is highly recommended.

Table 11.6 in Altman (1991) contains an example on ventricular circumferential shortening velocity versus fasting blood glucose by Thuesen et al.

We used those data to illustrate subsetting and use them again in the chapter on correlation and regression. They are among the built-in datasets in the ISwR package and available via data(thuesen), but the point here is to show how to read them from a plain-text file.

Let's assume that the data are contained in the file thuesen.txt, which looks as follows:

```
blood.glucose    short.velocity
15.3             1.76
10.8             1.34
8.1              1.27
19.5             1.47
7.2              1.27
5.3              1.49
9.3              1.31
11.1             1.09
7.5              1.18
12.2             1.22
6.7              1.25
5.2              1.19
19.0             1.95
15.1             1.28
6.7              1.52
8.6              NA
4.2              1.12
10.3             1.37
12.5             1.19
16.1             1.05
13.3             1.32
4.9              1.03
8.8              1.12
9.5              1.70
```

To enter the data into the file, you could start up Windows' NotePad (usually found via the Start button and Programs/Accessories) or a similar text editor and simply type the data as shown. Unix/Linux users should just use a standard editor like emacs or vi. Notice that the columns are separated by an arbitrary number of blanks and that NA represents a missing value.

Finally, you would save the data to a text file. Notice that some programs like Word or WordPad require special actions in order to save as text. Their normal save format is difficult to read from other programs.

Assuming further that the file is in the ISwR folder on the N: drive, the data can be read using

```
> thuesen <- read.table("N:/ISwR/thuesen.txt",header=T)
```

1.6 Data entry 41

Notice `header=T` specifying that the first line is a header containing the names of variables contained in the file. Also note that you use forward slashes (/), not backslashes (\), in the filename, even on a Windows system.

The reason for avoiding backslashes in filenames is that the symbol is used as an "escape character" for specifying characters that could not normally be entered in a text string: \n means the newline character; \ " means that the string should contain a "; etc. The backslash itself is written \\ so we could also have used `N:\\ISwR\\thuesen.txt`.

Variants of the `read.table` function are `read.csv` and `read.csv2`. The former assumes that fields are separated by a comma instead of whitespace, and the latter assumes that they are separated by semicolons but use a comma as the decimal point (some programs generate this format when running in European locales). In these formats an empty field is allowed to represent a missing value. Both have `header=T` as the default. Further variants are `read.delim` and `read.delim2` for reading delimited files (by default, the delimiter is the Tab character).

The result is a data frame, which is put into the variable `thuesen` and looks as follows:

```
> thuesen
   blood.glucose short.velocity
1           15.3           1.76
2           10.8           1.34
3            8.1           1.27
4           19.5           1.47
5            7.2           1.27
6            5.3           1.49
7            9.3           1.31
8           11.1           1.09
9            7.5           1.18
10          12.2           1.22
11           6.7           1.25
12           5.2           1.19
13          19.0           1.95
14          15.1           1.28
15           6.7           1.52
16           8.6             NA
17           4.2           1.12
18          10.3           1.37
19          12.5           1.19
20          16.1           1.05
21          13.3           1.32
22           4.9           1.03
23           8.8           1.12
24           9.5           1.70
```

The read.table function knows a couple of other tricks. It autodetects whether a vector is text or numeric and converts it to a factor in the former case (but makes no attempt to recognize numerically coded factors). Furthermore, it will recognize the case where the first line is one item shorter than the rest and will interpret that layout to imply that the first line contains a header and the first value on all subsequent lines is a row label — that is, exactly the layout generated when printing a data frame as in the above example. It is also possible to specify which strings represent missing values via the na.strings argument.

1.6.2   The data editor

R provides two ways of editing data interactively. One allows you to edit numeric variables in the workspace using the data.entry function, and the other lets you edit data frames. Both use the same spreadsheet-like interface. We only discuss the data frame editor here. The interface is a bit rough but quite useful for small data sets.

To edit a data frame you can use the edit function:

```
> data(airquality)
> aq <- edit(airquality)
```

This brings up a spreadsheet-like editor with a column for each variable in the data frame. The airquality data set is built into R; see help(airquality) for its contents. Inside the editor, you can move around with the mouse or the cursor keys and edit the current cell by typing in data. The type of variable can be switched between real (numeric) and character (factor) by clicking on the column header, and the name of the variable can be changed similarly.

Note that there is (as of R 1.5.0) no way to delete rows and columns and that new data can be entered only at the end.

When you close the data editor, the edited data frame is assigned to aq. The original airquality is left intact. Alternatively, if you do not mind overwriting the original data frame, you could have used

```
> fix(airquality)
```

In the airquality data, it may be useful to change the Month variable to a factor by switching its mode to character. This gives a factor with label names "5" to "9", but more meaningful names may be assigned after editing with

```
> levels(airquality$Month) <- c("May","June","July",
+ "August","September")
```

Alternatively, `levels(airquality$Month) <- month.name[5:9]` could have been used.

To enter data into a blank data frame, use

```
> dd <- data.frame()
> fix(dd)
```

An alternative would be `dd <- edit(data.frame())`, which works fine, except that beginners tend to re-execute the command when they need to edit `dd`, which of course destroys all data. It is necessary in either case to start with an empty data frame since by default `edit` expects you to want to edit a user-defined function and would bring up a text editor if you started it as `edit()`.

1.6.3 Interfacing to other programs

Sometimes you will want to move data between R and other statistical packages or spreadsheets. A simple fallback approach is to request the package in question to export data as a text file of some sort and use one of `read.table`, `read.csv`, `read.csv2`, `read.delim`, and `read.delim2` to read the data as an R data frame. Unfortunately, each program seems to have its own idiosyncrasies, but the read functions in R are flexible enough to cope.

The `foreign` package is one of the packages labeled "recommended" and should therefore be available with binary distributions of R. It contains routines to read files from SPSS (`.sav` format), SAS (export libraries), Epi-Info (`.rec`), Stata, Minitab, and some S-PLUS version 3 dump files.

Unix/Linux users sometimes find themselves with data sets written on Windows machines. The `foreign` package will work there as well for those formats that it supports. Notice that ordinary SAS data sets are not among the supported formats. These have to be converted to export libraries on the originating system. Data that have been entered into Microsoft Excel spreadsheets are most conveniently extracted using compatible applications like Sun StarOffice.

For data stored in databases, there exists a number of interface packages on CRAN. Of particular interest on Windows and with some Unix databases is the RODBC package, because you can set up ODBC connections to data stored by common applications including Excel and Access. Some Unix databases, e.g., PostgreSQL, also allow ODBC connections.

For up-to-date information on these matters, consult the "R Data Import/Export" manual that comes with the system.

1.7 Exercises

1.1 What is the most convenient way to insert a value between two elements of a vector at a given position?

1.2 How would you check whether two vectors are the same, if they may contain missing (NA) values? (Use of the identical function is considered cheating!)

1.3 If x is a factor with n levels and y is a length n vector, what happens if you compute y[x]?

1.4 The cut function is used to create a factor from a numeric vector. Look up the details on its help page. Create a factor in which the blood.glucose variable in the thuesen data is divided into the intervals (4, 7], (7, 9], (9, 12], and (12, 20]. Change the level names to "low", "intermediate", "high", and "very high".

1.5 Write the logical expression to use to extract girls between 7 and 14 years of age in the juul data set.

1.6 What happens if you change the levels of a factor (with levels) and give the same value to two or more levels?

1.7 Use sapply to simulate the result of taking the mean of 100 random numbers from the normal distribution for 10 independent samples.

1.8 Write the built-in data set thuesen to a Tab-separated text file with write.table. View it with a text editor (depending on your system). Change the NA value to . and read the changed file back into R with a suitable command.

Also try importing the data into other applications of your choice and exporting them to a new file after editing. You may have to remove row names to make this work.

2
Probability and distributions

The concepts of randomness and probability are central to statistics. It is an empirical fact that most experiments and investigations are not perfectly reproducible. The degree of irreproducibility may vary: Some physical experiments may yield data that are accurate to many decimal places, whereas data on biological systems are typically much less reliable. However, the view of data as something coming from a statistical distribution is vital to understanding statistical methods.

In this section we outline the basic ideas of probability and the functions that R has for random sampling and handling of theoretical distributions.

2.1 Random sampling

Much of the earliest work in probability theory was about games and gambling issues, based on symmetry considerations. The basic notion then is that of a random sample, dealing from a well-shuffled pack of cards or picking numbered balls from a well-stirred urn.

In R you can simulate these situations with the sample function. If you want to pick five numbers at random from the set 1:40, then you can write

```
> sample(1:40,5)
```

```
[1]   4 30 28 40 13
```

The first argument (x) is a vector of values to be sampled and the second
(size) is the sample size. Actually, sample(40,5) would suffice since
a single number is interpreted to represent the length of a sequence of
integers.

Notice that the default behaviour of sample is *sampling without replacement*. That is, the samples will not contain the same number twice, and
size can obviously not be bigger than the length of the vector to be sampled. If you want sampling with replacement, then you need to add the
argument replace=TRUE.

Sampling with replacement is suitable for modelling coin tosses or throws
of a die. So, for instance, to simulate 10 coin tosses we could write

```
> sample(c("H","T"), 10, replace=T)
 [1] "T" "T" "T" "T" "T" "H" "H" "T" "H" "T"
```

In fair coin-tossing, the probability of heads should equal the probability
of tails, but the idea of a random event is not restricted to symmetric cases.
It could be equally well applied to other cases such as the successful outcome of a surgical procedure. Hopefully, there would be a better than 50%
chance of this. You can simulate data with nonequal probabilities for the
outcomes (say, a 90% chance of success) by using the prob argument to
sample, as in

```
> sample(c("succ", "fail"), 10, replace=T, prob=c(0.9, 0.1))
 [1] "succ" "succ" "succ" "succ" "succ" "succ" "succ" "succ"
 [9] "succ" "succ"
```

This may not be the best way to generate such a sample, though. See the
later discussion of the binomial distribution.

2.2   Probability calculations and combinatorics

Let's return to the case of sampling without replacement, specifically
sample(1:40, 5). The probability of obtaining a given number as the
first one of the sample should be 1/40, the next one 1/39, and so forth. The
probability of a given sample should then be $1/(40 \times 39 \times 38 \times 37 \times 36)$.
Or, in R, use the prod function, which calculates the product of a vector
of numbers

```
> 1/prod(40:36)
[1] 1.266449e-08
```

However, notice that this is the probability of getting given numbers in a given order. If this were a Lotto-like game, then you would rather be interested in the probability of guessing a given *set* of five numbers correctly. Thus you need also to include the cases that give the same numbers in a different order. Since obviously the probability of each such case is going to be the same, all we need to do is to figure out how many such cases there are and multiply by that. There are 5 possibilities for the first number, and for each of these there are 4 possibilities for the second, and so forth; that is, the number is $5 \times 4 \times 3 \times 2 \times 1$. This number is also written as 5! (5 *factorial*). So the probability of a "winning Lotto coupon" would be

```
> prod(5:1)/prod(40:36)
[1] 1.519738e-06
```

There is another way of arriving at the same result. Notice that since the actual set of numbers is immaterial, all sets of five numbers must have the same probability. So all we need to do is to calculate the number of ways to choose 5 numbers out of 40. This is denoted

$$\binom{40}{5} = \frac{40!}{5!35!} = 658008$$

In R the choose function can be used to calculate this number, and the probability is thus

```
> 1/choose(40,5)
[1] 1.519738e-06
```

2.3 Discrete distributions

When looking at independent replications of a binary experiment, you would not usually be interested in whether each case is a success or a failure, but rather in the total number of successes (or failures). Obviously, this number is random, since it depends on the individual random outcomes, and it is consequently called a *random variable*. In this case it is a discrete-valued random variable that can take values in $0, 1, \ldots, n$, where n is the number of replications. Continuous random variables are encountered later.

A random variable X has a *probability distribution* that can be described using *point probabilities* $f(x) = P(X = x)$ or the *cumulative distribution function* $F(x) = P(X \le x)$. In the case at hand, the distribution can be worked out as having the point probabilities

$$f(x) = \binom{n}{x} p^x (1 - p)^{n-x}$$

This is known as the *binomial distribution*, and the $\binom{n}{x}$ are known as *binomial coefficients*. The parameter p is the probability of a successful outcome in an individual trial. A graph of the point probabilities of the binomial distribution appears in Figure 2.2 ahead.

We delay describing the R functions related to the binomial distribution until we have discussed continuous distributions so that we can present the conventions in a unified manner.

Many other distributions can be derived from simple probability models. For instance, the *geometric distribution* is similar to the binomial distribution but records the number of failures that occur before the first success.

2.4   Continuous distributions

Some data arise from measurements on an essentially continuous scale, for instance temperature, concentrations, etc. In practice they will be recorded to a finite precision, but it is useful to disregard this in the modelling. Such measurements will usually have a component of random variation, which makes them less than perfectly reproducible. However, these random fluctuations will tend to follow patterns; typically they will cluster around a central value with large deviations being more rare than smaller ones.

In order to model continuous data we need to define random variables that can obtain the value of any real number. Because there are infinitely many numbers infinitely close, the probability of any particular value will be zero so there is no such thing as a point probability as for discrete-valued random variables. Instead we have the concept of a *density*: This is the infinitesimal probability of hitting a small region around x divided by the size of the region. The cumulative distribution function can be defined as before, and we have the relation

$$F(x) = \int_{-\infty}^{x} f(x)\, dx$$

There are a number of standard distributions that come up in statistical theory and are available in R. It makes little sense to describe them in detail here except for a couple of examples.

The *uniform distribution* has a constant density over a specified interval (by default $[0, 1]$).

The *normal distribution* (also known as the *Gaussian distribution*) has density

$$f(x) = \frac{1}{\sqrt{2\pi}\sigma} \exp(-\frac{(x - \mu)^2}{2\sigma^2})$$

depending on its mean μ and standard deviation σ. The normal distribution has a characteristic bell shape (Figure 2.1), and modifying μ and σ simply translates and widens the distribution. It is a standard building block in statistical models, where it is commonly used to describe error variation. It also comes up as an approximating distribution in several contexts; for instance, the binomial distribution for large sample sizes can be well approximated by a suitably scaled normal distribution.

2.5 The built-in distributions in R

The standard distributions that turn up in connection with model building and statistical tests have been built into R, and it can therefore completely replace traditional statistical tables. Here we look only at the normal distribution and the binomial distribution, but other distributions follow exactly the same pattern.

Four fundamental items can be calculated for a statistical distribution:

- Density or point probability
- Cumulated probability, distribution function
- Quantiles
- Pseudo-random numbers

For all distributions implemented in R, there is a function for each of the four items listed above. For example, for the normal distribution, these are named dnorm, pnorm, qnorm, and rnorm, respectively (*d*ensity, *p*robability, *q*uantile, and *r*andom).

2.5.1 Densities

The density for a continuous distribution is a measure of the relative probability of "getting a value close to x". The probability of getting a value in a particular interval is the area under the corresponding part of the curve.

For discrete distributions, the term "density" is used for the point probability — the probability of getting exactly the value x. Technically, this is
correct: It is a density with respect to counting measure.

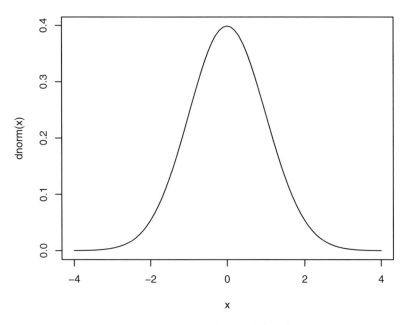

Figure 2.1. Density of normal distribution.

The density function is likely the one of the four function types that is least
used in practice, but if, for instance it is desired to draw the well-known
bell curve of the normal distribution, then it can be done like this:

```
> x <- seq(-4,4,0.1)
> plot(x,dnorm(x),type="l")
```

(Notice that this is the letter 'l', not the digit '1').

The function seq (see p. 13) is used to generate equidistant values, here
from -4 to 4 in steps of 0.1, that is $(-4.0, -3.9, -3.8, \ldots, 3.9, 4.0)$. The use
of type="l" as an argument to plot causes the function to draw lines
between the points rather than plotting the points themselves.

An alternative way of creating the plot is to use curve as follows:

```
> curve(dnorm(x), from=-4, to=4)
```

This is often a more convenient way of making graphs, but it does require that the *y*-values can be expressed as a simple functional expression in *x*.

For discrete distributions, where variables can take on only distinct values, it is preferable to draw a pin diagram, here for the binomial distribution with $n = 50$ and $p = 0.33$ (Figure 2.2):

```
> x <- 0:50
> plot(x,dbinom(x,size=50,prob=.33),type="h")
```

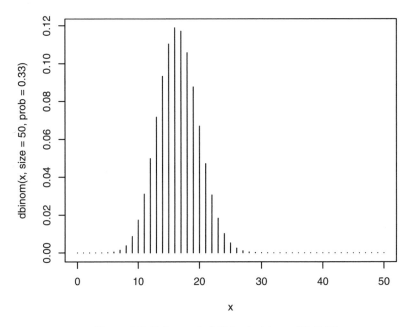

Figure 2.2. Point probabilities in binom(50, 0.33).

Notice that there are three arguments to the "d-function" this time. In addition to *x*, you have to specify the number of trials *n* and the probability parameter *p*. The distribution drawn corresponds to, for example, the number of 5s or 6s in 50 throws of a symmetrical die. Actually, dnorm also takes more than one argument, namely the mean and standard deviation, but they have default values of 0 and 1, respectively, since most often it is the standard normal distribution that is requested.

The form 0:50 is a short version of seq(0,50,1): the whole numbers from 0 to 50 (cf. p. 13). It is type="h" (as in *h*istogram-like) that causes the pins to be drawn.

2.5.2   Cumulative distribution functions

The cumulative distribution function describes the probability of "hitting" *x* or less in a given distribution. The corresponding R functions begin with a 'p' (for probability) by convention.

Just as you can plot densities, you can of course also plot cumulative distribution functions, but that is usually not very informative. More often, actual numbers are desired. Say that it is known that some biochemical measure in healthy individuals is well described by a normal distribution with a mean of 132 and a standard deviation of 13. Then, if a patient has a value of 160, there is

```
> 1-pnorm(160,mean=132,sd=13)
[1] 0.01562612
```

or only about 1.5% of the general population that has that value or higher. The function pnorm returns the probability of getting a value smaller than its first argument in a normal distribution with the given mean and standard deviation.

Another typical application occurs in connection with statistical tests. Consider a simple sign test: Twenty patients are given two treatments each (blindly and in randomized order) and then asked whether treatment A or B worked better. It turned out that 16 patients liked A better. The question is then whether this can be taken as sufficient evidence that A actually is the better treatment, or whether the outcome might as well have happened by chance even if the treatments were equally good. If there was no difference between the two treatments, then we would expect that the number of people favouring treatment A to be binomially distributed with $p = 0.5$ and $n = 20$. How (im)probable would it then be to obtain what we have observed? Like in the normal distribution, we need a tail probability, and the immediate guess might be to look at

```
> pbinom(16,size=20,prob=.5)
[1] 0.9987116
```

and subtract it from 1 to get the upper tail — but this would be an error! What we need is the probability of *the observed or more extreme* and pbinom is giving the probability of 16 or less. We need to use "15 or less" instead.

```
> 1-pbinom(15,size=20,prob=.5)
[1] 0.005908966
```

If you want a two-tailed test because you have no prior idea about which treatment is better, then you will have to add the probability of obtaining equally extreme results in the opposite direction. In the present case, that means the probability that 4 or fewer people prefer A, giving a total probability of

```
> 1-pbinom(15,20,.5)+pbinom(4,20,.5)
[1] 0.01181793
```

(which is obviously exactly twice the one-tailed probability).

As can be seen from the last command, it is not strictly necessary to use the `size` and `prob` keywords as long as the arguments are given in the right order (positional matching; Section 1.2.2).

It is quite confusing to keep track of whether or not the observation itself needs to be counted. Fortunately, the function `binom.test` keeps track of such formalities and performs the correct binomial test. This is further discussed in Chapter 7.

2.5.3 Quantiles

The quantile function is the inverse of the cumulative distribution function. The p-quantile is the value with the property that there is probability p of getting a value less than or equal to it. The median is by definition the 50% quantile.

Some details concerning the definition in the case of discontinuous distributions are glossed over here. You can fairly easily deduce the behaviour by experimenting with the R functions.

Tables of statistical distributions are almost all given in terms of quantiles. For a fixed set of probabilities, the table shows the boundary that a test statistic must cross in order to be considered significant at that level. This is purely for operational reasons; it is almost superfluous when you have the option of computing p exactly.

Theoretical quantiles are commonly used for the calculation of confidence intervals and for power calculations in connection with designing and dimensioning experiments (see Chapter 8). A simple example of a confidence interval can be given here (see also Chapter 4).

If we have n normally distributed observations with the same mean μ and standard deviation σ, then it is known that the average $\bar{x}$ is normally distributed around μ with standard deviation $\sigma/\sqrt{n}$. A 95% confidence interval for μ can be obtained as

$$\bar{x} + \sigma/\sqrt{n} \times N_{0.025} \leq \mu \leq \bar{x} + \sigma/\sqrt{n} \times N_{0.975}$$

where $N_{0.025}$ is the 2.5% quantile in the normal distribution. If $\sigma = 12$ and we have measured $n = 5$ persons and found an average of $\bar{x} = 83$, then we can compute the relevant quantities as follows ("sem" means *standard error of the mean*):

```
> xbar <- 83
> sigma <- 12
> n <- 5
> sem <- sigma/sqrt(n)
> sem
[1] 5.366563
> xbar + sem * qnorm(0.025)
[1] 72.48173
> xbar + sem * qnorm(0.975)
[1] 93.51827
```

and thus find a 95% confidence interval for μ going from 72.48 to 93.52.

Since it is known that the normal distribution is symmetric, so that $N_{0.025} = -N_{0.975}$, it is common to write the formula for the confidence interval as $\bar{x} \pm \sigma/\sqrt{n} \times N_{0.975}$. The quantile itself is often written $\Phi^{-1}(0.975)$, where Φ is standard notation for the cumulative distribution function of the normal distribution (pnorm).

Another application of quantiles is in connection with Q–Q plots (see Section 3.2.3), which can be used to assess whether a set of data can reasonably be assumed to come from a given distribution.

2.5.4 Random numbers

To many people it sounds like a contradiction in terms to generate random numbers on a computer, since its results are supposed to be predictable and reproducible. What is in fact possible is to generate sequences of "pseudo-random" numbers, which for practical purposes behave *as if* they were drawn randomly.

Here random numbers are used to give the reader a feeling for the way in which randomness affects the quantities that can be calculated from a set of data. In professional statistics they are used to create simulated data sets in order to study the accuracy of mathematical approximations and the effect of assumptions being violated.

The use of the functions that generate random numbers is straightforward. The first argument specifies the number of random numbers to compute, and the subsequent arguments are similar to those for other functions related to the same distributions. For instance,

```
> rnorm(10)
 [1] -0.2996466 -0.1718510 -0.1955634  1.2280843 -2.6074190
 [6] -0.2999453 -0.4655102 -1.5680666  1.2545876 -1.8028839
```

```
> rnorm(10)
 [1]  1.7082495   0.1432875  -1.0271750  -0.9246647   0.6402383
 [6]  0.7201677  -0.3071239   1.2090712   0.8699669   0.5882753
> rnorm(10,mean=7,sd=5)
 [1]  8.934983   8.611855   4.675578   3.670129   4.223117   5.484290
 [7] 12.141946   8.057541  -2.893164  13.590586
> rbinom(10,size=20,prob=.5)
 [1] 12 11 10  8 11  8 11  8  8 13
```

2.6 Exercises

2.1 Calculate the probability for each of the following events: (a) A standard normally distributed variable is larger than 3. (b) A normally distributed variable with mean 35 and standard deviation 6 is larger than 42. (c) Getting 10 out of 10 successes in a binomial distribution with probability 0.8. (d) $X < 0.9$ when X has the standard uniform distribution. (e) $X > 6.5$ in a χ^2 distribution with 2 degrees of freedom.

2.2 It is well known that 5% of the normal distribution lies outside an interval approximately $\pm 2s$ about the mean. Where are the limits corresponding to 1%, 5‰, and 1‰? What is the position of the quartiles measured in standard deviation units?

2.3 For a disease known to have a postoperative complication frequency of 20%, a surgeon suggests a new procedure. He tests it on 10 patients and there are no complications. What is the probability of operating on 10 patients successfully with the traditional method?

2.4 Simulated coin-tossing is probably better done using `rbinom` than using `sample`. Explain how.

3

Descriptive statistics and graphics

Before going into the actual statistical modelling and analysis of a data set, it is often useful to make some simple characterizations of the data in terms of summary statistics and graphics.

3.1 Summary statistics for a single group

It is easy to calculate simple summary statistics with R. Here is how to calculate the mean, standard deviation, variance, and median.

```
> x <- rnorm(50)
> mean(x)
[1] 0.03301363
> sd(x)
[1] 1.069454
> var(x)
[1] 1.143731
> median(x)
[1] -0.08682795
```

Notice that the example starts with the generation of an artificial data vector x of 50 normally distributed observations. It is used in examples throughout this section. When reproducing the examples, you will not get exactly the same results since your random numbers will differ.

Empirical quantiles may be obtained with the function `quantile`, like this:

```
> quantile(x)
         0%         25%         50%         75%        100%
-2.60741896 -0.54495849 -0.08682795  0.70018536  2.98872414
```

As you see, by default you get the minimum, the maximum, and the three *quartiles* — the 0.25, 0.50, and 0.75 quantiles, so named because they correspond to a division into four parts. Similarly, we have *deciles* for $0.1, 0.2, \ldots, 0.9$, and *centiles* or *percentiles*. The difference between the first and third quartiles is called the *interquartile range* (IQR) and is sometimes used as a robust alternative to the standard deviation.

It is also possible to obtain other quantiles; this is done by adding an argument containing the desired percentage points. This, for example, is how to get the deciles:

```
> pvec <- seq(0,1,0.1)
> pvec
 [1] 0.0 0.1 0.2 0.3 0.4 0.5 0.6 0.7 0.8 0.9 1.0
> quantile(x,pvec)
         0%         10%         20%         30%         40%
-2.60741896 -1.07746896 -0.70409272 -0.46507213 -0.29976610
        50%         60%         70%         80%         90%
-0.08682795  0.19436950  0.49060129  0.90165137  1.31873981
       100%
 2.98872414
```

Beware that there are several possible definitions of empirical quantiles. The one R uses is based on a sum polygon where the ith ranking observation is the $(i-1)/(n-1)$ quantile and intermediate quantiles are obtained by linear interpolation. It sometimes confuses students that in a sample of 10 there will be 3 observations below the first quartile with this definition.

If there are missing values in data, things become a bit more complicated. For illustration, we use the following example:

The data set `juul` contains variables from an investigation performed by Anders Juul (Rigshospitalet, Department for Growth and Reproduction) concerning serum IGF-I (insulin-like growth factor) in a group of healthy humans, primarily school children. The data set is contained in the `ISwR` package and contains a number of variables, of which we only use `igf1` (serum IGF-I) for now, but later in the chapter we also use `tanner` (Tanner stage of puberty, a classification into five groups, based on appearance of primary and secondary sexual characteristics), `sex`, and `menarche` (indicating whether or not a girl has had her first period).

Attempting to calculate the mean of igf1 reveals a problem.

```
> data(juul)
> attach(juul)
>  mean(igf1)
[1] NA
```

R will not skip missing values unless explicitly requested to do so. The mean of a vector with an unknown value is unknown. However, you can give the na.rm argument (*not* available, *rem*ove) to request that missing values be removed:

```
> mean(igf1,na.rm=T)
[1] 340.168
```

There is one slightly annoying exception: The length function will not understand na.rm, so we cannot use it to count the number of nonmissing measurements of igf1. However, you can use

```
> sum(!is.na(igf1))
[1] 1018
```

The above construction uses the fact that if logical values are used in arithmetic, then TRUE is converted to 1 and FALSE to 0.

A nice summary display of a numeric variable is obtained from the summary function:

```
> summary(igf1)
   Min. 1st Qu.  Median    Mean 3rd Qu.    Max.   NA's
   25.0   202.3   313.5   340.2   462.8   915.0   321.0
```

The 1st Qu. and 3rd Qu. refer to the empirical quartiles (0.25 and 0.75 quantiles).

In fact, it is possible to summarize an entire data frame with

```
> summary(juul)
      age              menarche             sex
 Min.    : 0.170   Min.    : 1.000   Min.    :1.000
 1st Qu.: 9.053   1st Qu.: 1.000   1st Qu.:1.000
 Median :12.560   Median : 1.000   Median :2.000
 Mean    :15.095   Mean    : 1.476   Mean    :1.534
 3rd Qu.:16.855   3rd Qu.: 2.000   3rd Qu.:2.000
 Max.    :83.000   Max.    : 2.000   Max.    :2.000
 NA's    : 5.000   NA's    :635.000   NA's    :5.000
      igf1             tanner             testvol
 Min.    : 25.0   Min.    : 1.000   Min.    : 1.000
 1st Qu.:202.2   1st Qu.: 1.000   1st Qu.: 1.000
 Median :313.5   Median : 2.000   Median : 3.000
 Mean    :340.2   Mean    : 2.640   Mean    : 7.896
```

```
3rd Qu.:462.8    3rd Qu.:   5.000    3rd Qu.:  15.000
Max.    :915.0    Max.    :  5.000    Max.    :  30.000
NA's    :321.0    NA's    :240.000    NA's    :859.000
```

Notice that this data set has menarche, sex, and tanner coded as numeric variables even though they are clearly categorical. This can be
mended as follows:

```
> detach(juul)
> juul$sex <- factor(juul$sex,labels=c("M","F"))
> juul$menarche <- factor(juul$menarche,labels=c("No","Yes"))
> juul$tanner <- factor(juul$tanner,
+                        labels=c("I","II","III","IV","V"))
> attach(juul)
> summary(juul)
      age           menarche     sex          igf1
 Min.    : 0.170    No  :369    M   :621    Min.    : 25.0
 1st Qu.: 9.053     Yes :335    F   :713    1st Qu.:202.2
 Median :12.560     NA's:635    NA's:  5    Median :313.5
 Mean    :15.095                            Mean    :340.2
 3rd Qu.:16.855                             3rd Qu.:462.8
 Max.    :83.000                            Max.    :915.0
 NA's    : 5.000                            NA's    :321.0
 tanner          testvol
 I   :515    Min.    :  1.000
 II  :103    1st Qu.:  1.000
 III : 72    Median :  3.000
 IV  : 81    Mean    :  7.896
 V   :328    3rd Qu.: 15.000
 NA's:240    Max.    : 30.000
             NA's    :859.000
```

Notice how the display changes for the factor variables. Note also that
juul was detached and reattached after the modification. This is because
modifying a data frame does not affect any attached version. It was not
strictly necessary to do it here, because summary works directly on the
data frame whether attached or not.

In the above the variables sex, menarche, and tanner were converted
to factors with suitable level names (in the raw data these are represented
using numeric codes). The syntax x <- factor(x,labels=...) is a
short form for x <- factor(x) followed by levels(x) <- .... The
converted variables were put back into the data frame juul replacing the
original sex, tanner, and menarche variables. We might also have used
the transform function:

```
> juul <- transform(juul,
+     sex=factor(sex,labels=c("M","F")),
+     menarche=factor(menarche,labels=c("No","Yes")),
+     tanner=factor(tanner,labels=c("I","II","III","IV","V")))
```

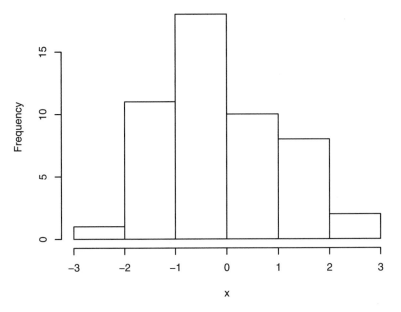

Figure 3.1. Histogram.

3.2 Graphical display of distributions

3.2.1 *Histograms*

You can get a reasonable impression of the shape of a distribution by drawing a histogram, that is, a count of how many observations fall within specified divisions ("bins") of the *x*-axis (Figure 3.1).

```
> hist(x)
```

By specifying breaks=n in the hist call, you get *approximately* n bars in the histogram since the algorithm tries to create "pretty" cutpoints. You can have full control over the interval divisions by specifying breaks as a vector, rather than a number. Altman (1991, pp. 25–26) contains an example of accident rates by age group. These are given as a count in age groups 0–4, 5–9, 10–15, 16, 17, 18–19, 20–24, 25–59, and 60–79 years of age. The data can be entered as follows:

```
> mid.age <- c(2.5,7.5,13,16.5,17.5,19,22.5,44.5,70.5)
> acc.count <- c(28,46,58,20,31,64,149,316,103)
> age.acc <- rep(mid.age,acc.count)
```

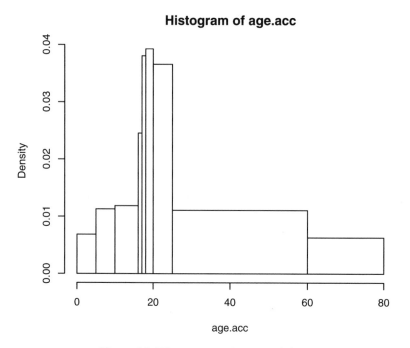

Figure 3.2. Histogram with unequal divisions.

```
> brk <- c(0,5,10,16,17,18,20,25,60,80)
> hist(age.acc,breaks=brk)
```

Here the first three lines generate pseudo-data from the table in the book. For each interval, the relevant number of "observations" is generated with an age set to the midpoint of the interval, that is, 28 2.5-year-olds, 46 7.5-year-olds, etc. Then a vector brk of cutpoints is defined (note that the extremes need to be included) and used as the breaks argument to hist, yielding Figure 3.2.

Notice that you automatically got the "correct" histogram where the *area* of a column is proportional to the number. The *y*-axis is in density units, that is, proportion of data per *x* unit, so that the total area of the histogram will be 1. If, for some reason, you wanted the (misleading) histogram where the column height is the raw number in each interval, then it can be specified using freq=T. For equidistant breakpoints, that is the default (because then you can see how many observations have gone into each column), but you can set freq=F to get densities displayed. This is really just a change of scale on the *y*-axis, but it has the advantage that it becomes possible to overlay the histogram with a corresponding theoretical density function.

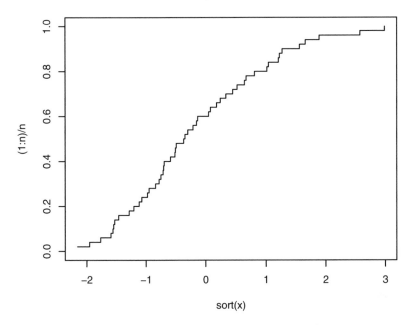

Figure 3.3. Empirical cumulative distribution function.

3.2.2   *Empirical cumulative distribution*

The empirical cumulative distribution function is defined as the fraction of data smaller than or equal to x. That is, if x is the kth smallest observation, then the proportion k/n of the data is smaller than or equal to x (7/10 if x is no. 7 of 10). The empirical cumulative distribution function can be plotted as follows, see Figure 3.3, where x is the simulated data vector from Section 3.1.

```
> n <- length(x)
> plot(sort(x),(1:n)/n,type="s",ylim=c(0,1))
```

The plotting parameter type="s" gives a *step* function where (x, y) is the left end of the steps and ylim is a vector of two elements specifying the extremes of the y-coordinates on the plot. Recall that c(...) is used to create vectors.

Some rather more elaborate displays of empirical cumulative distribution functions are available in the stepfun package.

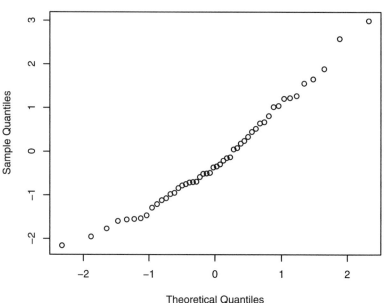

Figure 3.4. Probability plot, using qqnorm(x).

3.2.3 Q–Q plots

One purpose of calculating the empirical cumulative distribution function is to see whether data can be assumed normally distributed. For a better assessment, you might plot the k'th smallest observation against the expected value of the k'th smallest observation out of n in a standard normal distribution. The point is that in this way you would expect to obtain a straight line if data come from a normal distribution with *any* mean and standard deviation.

Creating such a plot is slightly complicated. Fortunately, there is a built-in function for doing it, qqnorm. The result of using it can be seen in Figure 3.4. You only have to write

```
> qqnorm(x)
```

As the title of the plot indicates, plots of this kind are also called "Q–Q plots" (quantile versus quantile). Notice that x and y are interchanged relative to the empirical c.d.f. — the observed values are now drawn along the y-axis. You should notice that with this convention, the distribution has heavy tails if the outer parts of the curve are steeper than the middle part.

Some readers will have been taught "probability plots" which are similar but have the axes interchanged. It can be argued that the way R draws the plot is the better one, since the theoretical quantiles are known in advance while the empirical quantiles depend on data. You would normally choose to draw fixed values horizontally and variable values vertically.

3.2.4 Boxplots

A "boxplot", or more descriptively a "box-and-whiskers plot", is a graphical summary of a distribution. Figure 3.5 shows boxplots for IgM and its logarithm; cf. the example on page 23 in Altman (1991).

Here is how a boxplot is drawn in R: The box in the middle indicates "hinges" (nearly quartiles, see the help page for boxplot.stats) and median. The lines ("whiskers") show the largest/smallest observation that falls within a distance of 1.5 times the box size from the nearest hinge. If any observations fall farther away, the additional points are considered "extreme" values and are shown separately.

The practicalities are these:

```
> data(IgM)
> par(mfrow=c(1,2))
> boxplot(IgM)
> boxplot(log(IgM))
> par(mfrow=c(1,1))
```

A layout with two plots side by side is specified using the mfrow graphical parameter. It should be read as "*multif*rame, *row*wise, 1×2 layout". Individual plots are organized in 1 row and 2 columns. As you might guess, there is also an mfcol parameter to plot columnwise. In a 2×2 layout the difference is whether plot no. 2 is drawn in the top right or bottom left corner.

Notice that it is necessary to reset the layout parameter to c(1,1) at the end, unless you also want two plots side by side subsequently.

3.3 Summary statistics by groups

When dealing with grouped data, you will often want to have various summary statistics computed within groups. For example, a table of

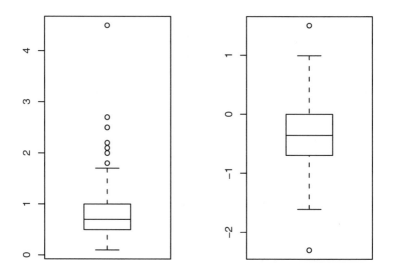

Figure 3.5. Boxplots for IgM and log IgM.

means and standard deviations. To this end you can use `tapply` (see Section 1.2.16). Here is an example concerning the folate concentration in red blood cells according to three types of ventilation during anesthesia (Altman, 1991, p. 208). We return to this example in Section 6.1, which also contains the explanation of the category names.

```
> data(red.cell.folate)
> attach(red.cell.folate)
> tapply(folate,ventilation,mean)
N2O+O2,24h  N2O+O2,op     O2,24h
  316.6250    256.4444   278.0000
```

The `tapply` call takes the `folate` variable, splits it according to `ventilation`, and computes the mean for each group. In the same way, standard deviations and number of observations in the groups can be computed.

```
> tapply(folate,ventilation,sd)
N2O+O2,24h  N2O+O2,op     O2,24h
  58.71709   37.12180   33.75648
> tapply(folate,ventilation,length)
N2O+O2,24h  N2O+O2,op     O2,24h
         8          9          5
```

Try something like this for a nicer display:

```
> xbar <- tapply(folate, ventilation, mean)
> s <- tapply(folate, ventilation, sd)
> n <- tapply(folate, ventilation, length)
> cbind(mean=xbar, std.dev=s, n=n)
                mean   std.dev n
N2O+O2,24h 316.6250 58.71709 8
N2O+O2,op  256.4444 37.12180 9
O2,24h     278.0000 33.75648 5
```

For the $juul$ data we might want the mean $igf1$ by $tanner$ group, but of course we run into the problem of missing values again:

```
> tapply(igf1, tanner, mean)
  I   II III  IV   V
 NA   NA  NA  NA   NA
```

We need to get $tapply$ to pass $na.rm=T$ as a parameter to $mean$ to make it exclude the missing values. This is achieved simply by passing it as an additional argument to $tapply$.

```
> tapply(igf1, tanner, mean, na.rm=T)
        I       II      III       IV        V
 207.4727 352.6714 483.2222 513.0172 465.3344
```

3.4 Graphics for grouped data

In dealing with grouped data it is important to be able not only to create plots for each group but also to be able to compare the plots between groups. In this section we review some general graphical techniques, allowing us to display similar plots for several groups on the same page. Some functions have specific features for displaying data from more than one group.

3.4.1 Histograms

We have already seen in Section 3.2.1 how to obtain a histogram simply by typing hist(x), where x is the variable containing the data. R will then choose a number of groups so that a reasonable number of data points falls in each bin, while at the same time ensuring that the cutpoints are "pretty" numbers on the x-axis.

It is also mentioned there that an alternative number of intervals can be set via the argument breaks, although you do not always get exactly

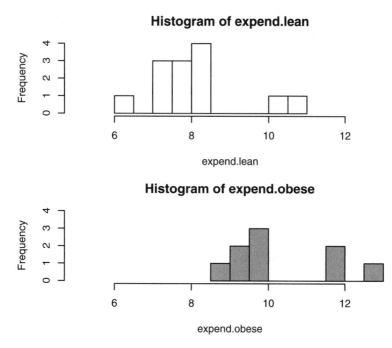

Figure 3.6. Histograms with refinements.

the number you asked for since R reserves the right to choose "pretty" column boundaries. For instance, multiples of 0.5 MJ are chosen in the following example using the energy data introduced in Section 1.2.14 on the 24-hour energy expenditure for two groups of women:

In this example some further techniques of general use are illustrated. The end result is seen in Figure 3.6, but first we must fetch the data:

```
> data(energy)
> attach(energy)
> expend.lean <- expend[stature=="lean"]
> expend.obese <- expend[stature=="obese"]
```

Notice how we separate the expend vector in the energy data frame into two vectors according to the value of the factor stature.

Then the actual plotting:

```
> par(mfrow=c(2,1))
> hist(expend.lean,breaks=10,xlim=c(5,13),ylim=c(0,4),col="white")
> hist(expend.obese,breaks=10,xlim=c(5,13),ylim=c(0,4),col="grey")
> par(mfrow=c(1,1))
```

We set par(mfrow=c(2,1)) to get the two histograms in the same plot. In the hist commands themselves, we used the breaks argument as already mentioned and col, whose effect should be rather obvious. We also used xlim and ylim to get the same x- and y-axes in the two plots. However, it is a coincidence that the columns have the same width.

A practical remark: When working with plots like the above where more than a single line of code is necessary to achieve the result, it gets cumbersome to use command recall in the R console window every time something needs modification. A better idea may be to start up a plain-text editor and cut and paste entire blocks of code from there. You might also take it as an incentive to start writing simple functions.

3.4.2   Parallel boxplots

You might want a set of boxplots from several groups in the same frame. boxplot can handle this, both when data are given in the form of separate vectors from each group and when data are in one long vector and a parallel vector or factor defines the grouping. To illustrate the latter we use the energy data introduced in Section 1.2.14

Figure 3.7 is created as follows:

```
> boxplot(expend ~ stature)
```

We could also have based the plot on the separate vectors expend.lean and expend.obese. In that case a syntax is used specifying the vectors as two separate arguments:

```
> boxplot(expend.lean,expend.obese)
```

The plot is not shown here, but the only difference lies in the labeling of the x-axis. There is also a third form where data are given as a single argument that is a list of vectors.

The bottom plot has been made using the complete expend vector and the grouping variable fstature.

Notation of the type y ~ x should be read "y described using x". This is the first example we see of a *model formula*. We see considerably more examples of model formulas when we get to regression analysis and analysis of variance in Chapters 5 and 6.

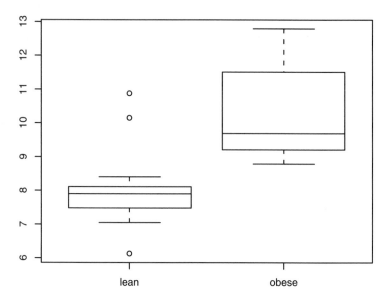

Figure 3.7. Parallel boxplot.

3.4.3 Stripcharts

The boxplots made in the preceding section show a "Laurel & Hardy" effect that is not really well founded in the data. The cause is that the interquartile range is quite a bit larger in one group than in the other, making the boxplot appear "fatter". With groups as small as these, the quartiles will be quite inaccurately determined, and it may therefore be more desirable to plot the raw data. If you were to do this by hand, you might draw a dot diagram where every number is marked with a dot on a number line. R's automated variant of this is the function stripchart. Four variants of stripcharts are seen in Figure 3.8.

The four plots were created as follows:

```
> opar <- par(mfrow=c(2,2),mex=0.8,mar=c(3,3,2,1)+.1)
> stripchart(expend~stature)
> stripchart(expend~stature,method="stack")
> stripchart(expend~stature,method="jitter")
> stripchart(expend~stature,"jitter",jitter=.03)
> par(opar)
```

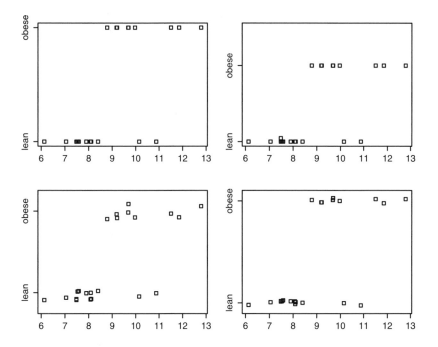

Figure 3.8. Stripcharts in four variations.

Notice that a little par magic was used to reduce the spacing between the four plots. The mex setting reduces the interline distance and mar reduces the number of lines that surround the plot region. This can be done for these plots since they have neither main title, subtitle, nor x and y labels. All the original values of the changed settings can be stored in a variable (here opar) and reestablished with par(opar).

The first plot is a standard stripchart, where the points are simply plotted on a line. The problem with this is that some points can become invisible because they are overplotted. This is why there is a method argument, which can be set to either "stack" or "jitter".

The former method stacks points with identical values, but it only does so if data are *completely identical*, so in the upper right plot, it is only the two replicates of 7.48 that get stacked, whereas 8.08, 8.09, and 8.11 are still plotted in almost the same spot.

The "jitter" method offsets all points a random amount vertically. The standard jittering on plot no. 3 (bottom left) is a bit large; it may be preferable to make it clearer that data are placed along a horizontal line. For that purpose, you can set jitter lower than the default of 0.1, which is done in the fourth plot.

In this example we have not bothered to specify data in several forms as we did for `boxplot` but used `expend~stature` throughout. We could also have written

```
stripchart(list(lean=expend.lean, obese=expend.obese))
```

but `stripchart(expend.lean, expend.obese)` cannot be used. (This is so that `method` can be used as a positional argument.)

3.5 Tables

Categorical data are usually described in the form of tables. This section outlines how you can create tables from your data and calculate relative frequencies.

3.5.1 Generating tables

We deal mainly with two-way tables. In the first example we enter a table directly, as is required for tables taken from a book or a journal article.

A two-way table needs to be in a *matrix* object (Section 1.2.6). Altman (1991, p. 242) contains an example on caffeine consumption by marital status among women giving birth. That table may be input as follows:

```
> caff.marital <- matrix(c(652,1537,598,242,36,46,38,21,218
+ ,327,106,67),
+ nrow=3,byrow=T)
> caff.marital
     [,1] [,2] [,3] [,4]
[1,]  652 1537  598  242
[2,]   36   46   38   21
[3,]  218  327  106   67
```

The `matrix` function needs an argument containing the table values as a single vector and also the number of rows in the argument `nrow`. By default, the values are entered columnwise; if rowwise entry is desired, then you need to specify `byrow=T`.

You might also give the number of columns instead of rows using `ncol`. If exactly one of `ncol` and `nrow` is given, R will compute the other one so that it fits the number of values. If both `ncol` and `nrow` are given and it does not fit the number of values, the values will be "recycled", which in some (other!) circumstances can be useful.

To get readable printouts, you can add row and column names to the matrices.

```
> colnames(caff.marital) <- c("0","1-150","151-300",">300")
> rownames(caff.marital) <- c("Married","Prev.married","Single")
> caff.marital
               0 1-150 151-300 >300
Married      652  1537     598  242
Prev.married  36    46      38   21
Single       218   327     106   67
```

In practice, the more frequent case is that you have a database of variables for each person in a data set. In that case, you should do the tabulation with `table`, `xtabs`, or `ftable`. These functions will generally work for tabulating numeric vectors as well as factor variables, but the latter will have their levels used for row and column names automatically. Hence, it is recommended to convert numerically coded categorical data into factors. The `table` function is the oldest and most basic of the three. The other two offer formula-based interfaces and better printing of multiway tables.

The data set `juul` is introduced on p. 58. Here we look at some other variables in that data set, namely `sex` and `menarche`; the latter indicates whether or not a girl has had her first period. We can generate some simple tables as follows:

```
> table(sex)
sex
  M   F
621 713
> table(sex,menarche)
   menarche
sex  No Yes
  M   0   0
  F 369 335
> table(menarche,tanner)
        tanner
menarche   I II III IV   V
     No  221 43  32 14   2
     Yes   1  1   5 26 202
```

Of course, the table of menarche versus sex is just a check on internal consistency of the data. The table of menarche versus Tanner stage of puberty is more interesting.

There are also tables with more than two sides, but not many simple statistical functions use them. Briefly, to tabulate such data just write, for example, `table(factor1, factor2, factor3)`. To input a table of cell counts, use the `array` function (an analog of `matrix`).

Like any matrix, a table can be transposed with the t function:

```
> t(caff.marital)
        Married Prev.married Single
0           652          36    218
1-150      1537          46    327
151-300     598          38    106
>300        242          21     67
```

For multiway tables, exchanging indices (generalized transposition) is done by aperm.

3.5.2 Marginal tables and relative frequency

It is often desired to compute marginal tables, that is, the sums of the counts along one or the other dimension of a table. Due to missing values, this might not coincide with just tabulating a single factor. This is done fairly easily using the apply function (Section 1.2.16), but there is also a simplified version called margin.table, described below.

First we need to generate the table itself:

```
> tanner.sex <- table(tanner,sex)
```

tanner.sex is an arbitrarily chosen variable name, which is used for the crosstable of tanner and sex.

```
> tanner.sex
        sex
tanner    M    F
    I   291  224
    II   55   48
    III  34   38
    IV   41   40
    V   124  204
```

Then we compute the marginal tables:

```
> margin.table(tanner.sex,1)
tanner
  I  II III  IV   V
515 103  72  81 328
> margin.table(tanner.sex,2)
sex
  M   F
545 554
```

The second argument to margin.table is the number of the marginal index: 1 and 2 give row and column totals, respectively.

Relative frequencies in a table are generally expressed as proportions of the row or column totals. Tables of relative frequencies can be constructed using prop.table, as follows:

```
> prop.table(tanner.sex,1)
     sex
tanner        M          F
   I    0.5650485 0.4349515
  II    0.5339806 0.4660194
 III    0.4722222 0.5277778
  IV    0.5061728 0.4938272
   V    0.3780488 0.6219512
```

Note that the *rows* (1st index) sum to 1. If a table of percentages is desired, just multiply the entire table by 100.

prop.table cannot be used to express the numbers relative to the grand total of the table, but you can of course always write

```
> tanner.sex/sum(tanner.sex)
     sex
tanner        M          F
   I    0.26478617 0.20382166
  II    0.05004550 0.04367607
 III    0.03093722 0.03457689
  IV    0.03730664 0.03639672
   V    0.11282985 0.18562329
```

The functions margin.table and prop.table also work on multiway tables — the margin argument can be a vector if the relevant margin has two or more dimensions.

3.6 Graphical display of tables

For presentation purposes, it may be desirable to display a graph rather than a table of counts or percentages. In this section the main methods for this are described.

3.6.1 Bar plots

Bar plots are made using barplot. This function takes an argument, which can be a vector or a matrix. The simplest variant goes as follows (Figure 3.9):

```
> total.caff <- margin.table(caff.marital,2)
```

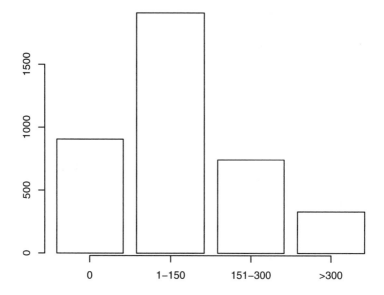

Figure 3.9. Simple `barplot` of total caffeine consumption.

```
> total.caff
     0   1-150 151-300    >300
   906    1910     742     330
> barplot(total.caff, col="white")
```

Without the `col="white"` argument, the plot comes out in colour, but this is not suitable for a black and white book illustration.

If the argument is a matrix, then `barplot` creates by default a "stacked bar plot", where the columns are partitioned according to the contributions from different rows of the table. If you want to place the row contributions beside each other instead, you can use the argument `beside=T`. A series of variants is found in Figure 3.10, which is constructed as follows:

```
> par(mfrow=c(2,2))
> barplot(caff.marital, col="white")
> barplot(t(caff.marital), col="white")
> barplot(t(caff.marital), col="white", beside=T)
> barplot(prop.table(t(caff.marital),2), col="white", beside=T)
> par(mfrow=c(1,1))
```

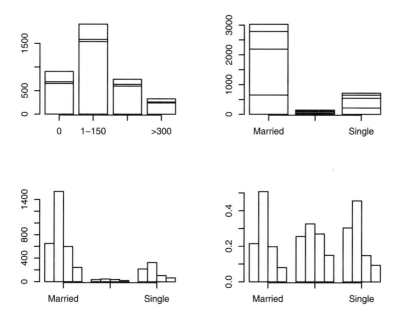

Figure 3.10. Four variants of `barplot` on a two-way table.

In the last three plots we switched rows and columns with the transposition function `t`. In the very last one the columns are expressed as proportions of the total number in the group. Thus, information is lost on the relative sizes of the marital status groups, but the group of previously married women (recall that the data set deals with women giving birth) is so small that it otherwise becomes almost impossible to compare their caffeine consumption profile with those of the other groups.

As usual, there is a multitude of ways to "prettify" the plots. Here is one possibility (Figure 3.11):

```
> barplot(prop.table(t(caff.marital),2),beside=T,
+ legend.text=colnames(caff.marital),
+ col=c("white","grey80","grey50","black"))
```

Notice that the legend overlaps the top of one of the columns. R is not designed to be able to find a clear area in which to place the legend. However, you can get full control of the legend's position if you insert it explicitly with the `legend` function. For that purpose, it will be helpful to use `locator()`, which allows you to click a mouse button over the plot and have the coordinates returned. See p. 174 for more about this.

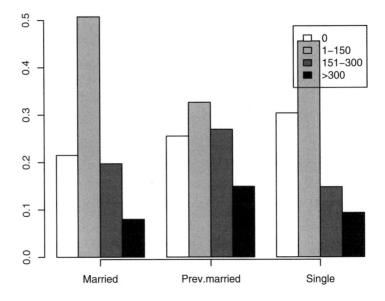

Figure 3.11. Bar plot with specified colours and legend.

3.6.2 Dotcharts

The Cleveland dotcharts, named after William S. Cleveland (1994), can be employed to study a table from both sides at the same time. They contain the same information as bar plots with beside=T but give quite a different visual impression. We content ourselves with a single example here (Figure 3.12):

```
> dotchart(t(caff.marital))
```

3.6.3 Pie charts

Pie charts are traditionally frowned upon by statisticians because they are so often used to make trivial data look impressive and are difficult to decode for the human mind. They very rarely contain information that would not have been at least as effectively conveyed in a bar plot. Once in a while they are useful, though, and it is no problem to get R to draw them. Here is a way to represent the table of caffeine consumption versus

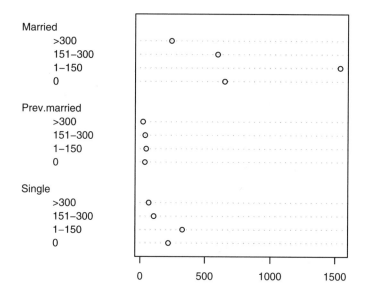

Figure 3.12. Dotchart of caffeine consumption.

marital status (Figure 3.13; see Section 3.4.3 for an explanation of the "par magic" used to reduce the space between the subplots):

```
> opar <- par(mfrow=c(2,2),mex=0.8, mar=c(1,1,2,1))
> slices <- c("white","grey80","grey50","black")
> pie(caff.marital["Married",], main="Married", col=slices)
> pie(caff.marital["Prev.married",],
+         main="Previously married", col=slices)
> pie(caff.marital["Single",], main="Single", col=slices)
> par(opar)
```

The col argument sets the colour of the pie slices.

There are more possibilities with piechart. The help page for pie contains an illustrative example concerning the distribution of pie sales(!) by pie type.

3.7   Exercises

3.1   Explore the possibilities for different kinds of line and point plots. Vary the plot symbol, line type, line width, and colour.

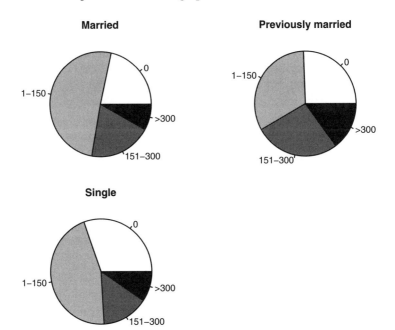

Figure 3.13. Pie charts of caffeine consumption according to marital status.

3.2 If you make a plot like `plot(rnorm(10),type="o")` with over-plotted lines and points, the lines will be visible inside the plotting symbols. How can this be avoided?

3.3 How can you overlay two `qqnorm` plots in the same plotting area? What goes wrong if you try to generate the plot using `type="l"` and how do you avoid that?

3.4 Plot a histogram for the `react` data set. Since these data are highly discretized, the histogram will be biased. Why? You may want to try `truehist` from the `MASS` package as a replacement.

3.5 Generate a sample vector z of 5 random numbers from the uniform distribution and make a line plot of `quantile(z,x)` as a function of x (use `curve`, for instance).

4

One- and two-sample tests

The rest of this book describes applications of R for actual statistical analysis. The focus to some extent shifts from explanation of the syntax to description of the output and of specific arguments to the relevant functions.

Some of the most basic statistical tests deal with comparing continuous data, either between two groups or against an a priori stipulated value. This is the topic for this chapter.

Two functions are introduced here, namely `t.test` and `wilcox.test` for t tests and Wilcoxon tests, respectively. Both can be applied to one- and two-sample problems as well as to paired data. Notice that the "two-sample Wilcoxon test" is the same as the one called the "Mann–Whitney test" in many textbooks.

4.1 One-sample t test

The t tests are based on an assumption that data come from the Normal distribution. In the one-sample case we thus have data $x_1, \ldots, x_n$ assumed independent realizations of random variables with distribution $N(\mu, \sigma^2)$, which denotes the normal distribution with mean μ and variance σ^2, and we wish to test the *null hypothesis* that $\mu = \mu_0$. We can estimate the parameters μ and σ by the empirical mean $\bar{x}$ and standard deviation s, although we must realize that we could never pinpoint their values exactly.

The key concept is that of the *standard error of the mean*, or SEM. This describes the variation of the average of n random values with mean μ and variance σ^2. This value is

$$\mathrm{SEM} = \sigma/\sqrt{n}$$

and means that if you were to repeat the entire experiment several times and calculate an average for each experiment, then these averages would follow a distribution that is narrower than that of the original distribution. The crucial point is that even based on a single sample, it is possible to calculate an empirical SEM as $s/\sqrt{n}$ using the empirical standard deviation of the sample. This value will tell us how far the observed mean may reasonably have strayed from its true value. For normally distributed data, the rule of thumb is that there is 95% probability of staying within $\mu \pm 2\sigma$ so we would expect that if μ_0 were the true mean, then $\bar{x}$ should be within 2 SEM's of it. Formally, you calculate

$$t = \frac{\bar{x} - \mu_0}{\mathrm{SEM}}$$

and see whether this falls within an *acceptance region*, outside which t should fall with probability equal to a specified *significance level*. This is often chosen as 5%, in which case the acceptance region is almost, but not exactly, the interval from -2 to 2.

In small samples it is necessary to correct for the fact that an empirical SEM is used and that the distribution of t therefore has somewhat "heavier tails" than the $N(0,1)$: Large deviations happen more frequently than in the normal distribution since they can result from normalizing with an SEM that is too small. The correct values for the acceptance region can be looked up as quantiles in the t distribution with $f = n - 1$ degrees of freedom.

If t falls outside the acceptance region, then we reject the null hypothesis at the chosen significance level. Alternatively (and equivalently), you can calculate the *p-value*, which is the probability of obtaining a value as numerically large as or larger than the observed t and reject the hypothesis if the p-value is less than the significance level.

Sometimes you have prior information on the direction of an effect: for instance, that all plausible mechanisms that would cause μ not to equal μ_0 would tend to make it bigger. In those cases, you may choose to reject the hypothesis only if t falls in the upper tail of the distribution. This is known as *testing against a one-sided alternative*. Since removing the lower tail from the rejection region effectively halves the significance level, a one-sided test at a given level will have a smaller cutoff point. Similarly, p-values are calculated as the probability of a larger value than the observed rather than a numerically larger one, effectively halving the *p*-value as long as the observed effect is in the stipulated direction. One-sided tests should be used with some care, preferably only when there is a clear statement of the intent to use them in the study protocol. Switching to a one-sided test to make an otherwise nonsignificant result significant could easily be regarded as dishonest.

This is an example concerning daily energy intake in kJ for 11 women (Altman, 1991, p. 183). First, the values are placed in a data vector:

```
> daily.intake <- c(5260,5470,5640,6180,6390,6515,
+ 6805,7515,7515,8230,8770)
```

Let's first look at some simple summary statistics, even though these are hardly necessary for such a small data set:

```
> mean(daily.intake)
[1] 6753.636
> sd(daily.intake)
[1] 1142.123
> quantile(daily.intake)
  0%  25%  50%  75% 100%
5260 5910 6515 7515 8770
```

You might wish to investigate whether the women's energy intake deviates systematically from a recommended value of 7725 kJ. Assuming that data comes from a normal distribution, the object is to test whether this distribution might have mean $\mu = 7725$. This is done with t.test, as follows:

```
> t.test(daily.intake,mu=7725)

        One Sample t-test

data:  daily.intake
t = -2.8208, df = 10, p-value = 0.01814
alternative hypothesis: true mean is not equal to 7725
95 percent confidence interval:
 5986.348 7520.925
sample estimates:
mean of x
 6753.636
```

This is an example of the exact same type as that used in the introductory Section 1.1.4. The description of the output is quite superficial there. Here it is explained more thoroughly.

The layout is common to many of the statistical standard tests, and a "dissection" is given in the following:

```
One Sample t-test
```

This should be self-explanatory. It is simply a description of the test that we have asked for. Notice that, by looking at the format of the function call, t.test has automatically found out that a one-sample test is desired.

```
data:   daily.intake
```

This tells us which data are being tested. Of course, this will be obvious, *unless* output has been separated from the command that generated it. This can happen, for example, when using the source function to read commands from an external file.

```
t = -2.8208, df = 10, p-value = 0.01814
```

This is where it begins to get interesting. We get the t statistic, the associated degrees of freedom, and the exact p-value. We do not need to use a table of the t distribution to look up which quantiles the t-value can be found between. You can immediately see that $p < 0.05$ and thus that (using the customary 5% level of significance) data deviate significantly from the hypothesis that the mean is 7725.

```
alternative hypothesis: true mean is not equal to 7725
```

This contains two important pieces of information: (a) the value we wanted to test whether the mean could be equal to (7725 kJ) and (b) that the test is two-sided ("not equal to").

```
95 percent confidence interval:
 5986.348 7520.925
```

This is a 95% confidence interval for the true mean, that is, the set of (hypothetical) mean values from which the data do not deviate significantly. It is based on inverting the t test by solving for the values of μ_0 that cause t to lie within its acceptance region. For a 95% confidence interval, the solution is

$$\bar{x} - t_{0.975}(f) \times s < \mu < \bar{x} + t_{0.975}(f) \times s$$

```
sample estimates:
mean of x
 6753.636
```

This final item is the observed mean, that is, the (point) estimate of the true mean.

The function t.test has a number of optional arguments, three of which are relevant in one-sample problems. We have already seen the use of mu to specify the mean value μ under the null hypothesis (default is mu=0). In addition, you can specify that a one-sided test is desired against alternatives greater than μ by using alternative="greater" or alternatives less than μ using alternative="less". The third item that can be specified is the *confidence level* used for the confidence intervals; you would write conf.level=0.99 to get a 99% interval.

Actually, it is often allowable to abbreviate a longish argument specification; for instance, it is sufficient to write alt="g" to get the test against greater alternatives.

4.2   Wilcoxon signed-rank test

The t tests are fairly robust against departures from the normal distribution especially in larger samples, but sometimes you wish to avoid making that assumption. To this end, the *distribution-free methods* are convenient. These are generally obtained by replacing data with corresponding order statistics.

For the one-sample Wilcoxon test, the procedure is to subtract the theoretical μ_0 and rank the differences according to their numerical value ignoring the sign, then calculate the sum of the positive or negative ranks. The point is that assuming only that the distribution is symmetric around μ_0, the test statistic corresponds to selecting each number from 1 to n with probability $1/2$ and calculating the sum. The distribution of the test statistic can be calculated exactly, at least in principle. It becomes computationally excessive in large samples, but the distribution is then very well approximated by a normal distribution.

Practical application of the Wilcoxon signed-rank test is done almost exactly as the t test:

```
> wilcox.test(daily.intake, mu=7725)
        Wilcoxon signed rank test with continuity correction

data:  daily.intake
V = 8, p-value = 0.0293
alternative hypothesis: true mu is not equal to 7725

Warning message:
Cannot compute exact p-value with ties in: wilcox.test.default(...
```

There is not quite as much output as from t.test, due to the fact that there is no such thing as a parameter estimate in a nonparametric test and therefore no confidence limits, etc. either. Actually, it is possible under some assumptions to define a location measure and calculate confidence intervals for it. See the help files for `wilcox.test` for details.

The relative merits of distribution-free (or *nonparametric*) versus parametric methods like the *t* test are a contentious issue. If the model assumptions of the parametric test are fulfilled, then it will be somewhat more efficient, on the order of 5% in large samples, although the difference can be larger in small samples. Notice, for instance, that unless the sample size is 6 or above, the signed-rank test simply cannot become significant at the 5% level. This is probably not too important, though; what is more important is that the apparent lack of assumptions for these tests sometimes misleads people into using them for data where the observations are not independent or where a comparison is biased by an important covariate.

The Wilcoxon tests are susceptible to the problem of *ties*, where several observations share the same value. In such cases you simply use the average of the tied ranks; for example, if there are four identical values corresponding to places 6 to 9, they will all be assigned the value 7.5. This is not a problem for the large-sample normal approximations, but the exact small-sample distributions become much more difficult to calculate and `wilcox.test` cannot do so.

The test statistic V is the sum of the positive ranks. In the example, the *p*-value is computed from the normal approximation because of the tie at 7515.

The function `wilcox.test` takes arguments `mu` and `alternative`, just like `t.test`. In addition, it has `correct` which turns a continuity correction on or off (default is "on", as seen from the output title; `correct=F` turns it off), and `exact`, which specifies whether exact tests should be calculated. Recall that "on/off" options like these are specified using logical values that can be either `TRUE` or `FALSE`.

4.3 Two-sample *t* test

The two-sample *t* test is used to test the hypothesis that two samples may be assumed to come from distributions with the same mean.

The theory for the two-sample *t* test is not very different in principle from that of the one-sample test. Data are now from two groups $x_{11}, \ldots, x_{1n_1}$ and $x_{21}, \ldots, x_{2n_2}$, which we assume are sampled from the normal distributions $N(\mu_1, \sigma_1^2)$ and $N(\mu_2, \sigma_2^2)$ and it is desired to test the null hypothesis $\mu_1 = \mu_2$. You then calculate

$$t = \frac{\bar{x}_2 - \bar{x}_1}{\text{SEDM}}$$

where the *standard error of difference of means* is

$$\text{SEDM} = \sqrt{\text{SEM}_1^2 + \text{SEM}_2^2}$$

There are two ways of calculating the SEDM depending on whether or not you assume that the two groups have the same variance. The "classical" approach is to assume that the variances are identical. With this approach you first calculate a pooled *s* based on the standard deviations from the two groups and plug that value into the SEM. Under the null hypothesis, the *t* value will follow a *t* distribution with $n_1 + n_2 - 2$ degrees of freedom.

An alternative procedure due to Welch is to calculate the SEMs from the separate group standard deviations s_1 and s_2. With this procedure, *t* is actually not *t*-distributed, but its distribution may be approximated by a *t* distribution with a number of degrees of freedom that can be calculated from s_1, s_2, and the group sizes. This is generally not an integer.

The Welch procedure is generally considered the safer one. Usually, the two procedures give very similar results unless both the group sizes and the standard deviations are very different.

We return to the daily energy expenditure data (see Section 1.2.14) and consider the problem of comparing energy expenditure between lean and obese women.

```
> data(energy)
> attach(energy)
> energy
    expend stature
1     9.21   obese
2     7.53    lean
3     7.48    lean
...
20    7.58    lean
21    9.19   obese
22    8.11    lean
```

Notice that the necessary information is contained in two parallel columns of a data frame. The factor `stature` contains the group and the numeric variable `expend` the energy expenditure in mega-Joules. R allows data in this format to be analyzed by `t.test` and `wilcox.test` using a model formula specification. An older format (still available) requires you to specify data from each group in a separate variable, but the newer format is much more convenient for data that are kept in data frames and is also more flexible if you later want to group the same response data according to other criteria.

The object is to see whether there is a shift in level between the two groups, so we apply a *t* test as follows:

```
> t.test(expend~stature)

        Welch Two Sample t-test

data:   expend by stature
t = -3.8555, df = 15.919, p-value = 0.001411
alternative hypothesis: true difference in means is not equal to 0
95 percent confidence interval:
 -3.459167 -1.004081
sample estimates:
 mean in group lean mean in group obese
          8.066154           10.297778
```

Notice the use of the tilde (~) operator to specify that expend is *described by* stature.

The output is not much different from that of the one-sample test. The confidence interval is for the *difference* in means and does not contain 0, which is in accordance with the *p*-value indicating a significant difference at the 5% level.

It is Welch's variant of the *t* test that is calculated by default. This is the test where you do not assume that the variance is the same in the two groups, which (among other things) results in the fractional degrees of freedom.

To get the usual (textbook) *t* test, you must specify that you are willing to assume that the variances are the same. This is done via the optional argument var.equal=T, that is:

```
> t.test(expend~stature, var.equal=T)

        Two Sample t-test

data:   expend by stature
t = -3.9456, df = 20, p-value = 0.000799
alternative hypothesis: true difference in means is not equal to 0
95 percent confidence interval:
 -3.411451 -1.051796
sample estimates:
 mean in group lean mean in group obese
          8.066154           10.297778
```

Notice that the degrees of freedom now has become a whole number, namely $13 + 9 - 2 = 20$. The *p*-value has dropped slightly (from 0.14% to 0.08%) and the confidence interval is a little narrower, but overall the changes are slight.

4.4   Comparison of variances

Even though it is possible in R to perform the two-sample t test without the assumption that the variances are the same, you may still be interested in testing that assumption, and R provides the var.test function for that purpose, implementing an F test on the ratio of the group variances. It is called the same way as t.test:

```
> var.test(expend~stature)

        F test to compare two variances

data:  expend by stature
F = 0.7844, num df = 12, denom df =  8, p-value = 0.6797
alternative hypothesis: true ratio of variances is not equal to 1
95 percent confidence interval:
 0.1867876 2.7547991
sample estimates:
ratio of variances
          0.784446
```

It is seen that the test is not significant, so there is no evidence against the assumption of the variances being identical. However, the confidence interval is very wide. For small data sets such as this one, the assumption of constant variance is largely a matter of belief. It may also be noted that this test is not robust against departures from a normal distribution. The ctest package which contains all the "classical tests", also has several alternative tests for variance homogeneity, each with its own assumptions, benefits, and drawbacks, but it would be excessive to discuss them at length.

Notice that the test is based on the assumption that the groups are independent. You should not apply this test to paired data.

4.5   Two-sample Wilcoxon test

You might prefer a nonparametric test if you doubt the normal distribution assumptions of the t test. The two-sample Wilcoxon test is based on replacing the data by their rank (without regard to grouping) and calculating the sum of the ranks in one group, thus reducing the problem to one of sampling n_1 values without replacement from the numbers 1 to $n_1 + n_2$.

This is done using wilcox.test, which behaves similarly to t.test:

```
> wilcox.test(expend~stature)
```

```
        Wilcoxon rank sum test with continuity correction

data:   expend by stature
W = 12, p-value = 0.002122
alternative hypothesis: true mu is not equal to 0

Warning message:
Cannot compute exact p-value with ties in: wilcox.test.default(...
```

The test statistic W is the sum of ranks in the first group minus its theoretical minimum (i.e., it is zero if all the smallest values fall in the first group). Some textbooks use a statistic that is the sum of ranks in the *smallest* group with no minimum correction, which is of course equivalent. Notice that, like in the one-sample example, we are having problems with ties and rely on the approximate normal distribution of W.

4.6   The paired t test

Paired tests are used when there are two measurements on the same experimental unit. Their theory is essentially based on taking differences and thus reducing the problem to that of a one-sample test. Notice, though, that it is implicitly assumed that such differences have a distribution that is independent of the level. A useful graphical check is to make a scatterplot of the pairs with the line of identity added, or to plot the difference against the average of the pair (sometimes called a *Bland–Altman plot*). If there seems to be a tendency for the dispersion to change with the level, then it may be useful to transform the data; frequently the standard deviation is proportional to the level, in which case a logarithmic transformation is useful.

The data on pre- and postmenstrual energy intake in a group of women are considered several times in Chapter 1 (and you may notice that the first column is identical to daily.intake, which was used in Section 4.1). There data are entered from the command line, but they are also available as a data set in the ISwR package:

```
> data(intake)
> attach(intake)
> intake
   pre post
1  5260 3910
2  5470 4220
3  5640 3885
4  6180 5160
5  6390 5645
6  6515 4680
```

```
7   6805 5265
8   7515 5975
9   7515 6790
10  8230 6900
11  8770 7335
```

The point is that the same 11 women are measured twice, so it makes sense to look at individual differences:

```
> post - pre
 [1] -1350 -1250 -1755 -1020  -745 -1835 -1540 -1540  -725 -1330
[11] -1435
```

— and it is immediately seen that they are all negative. All the women have a lower energy intake postmenstrually than premenstrually. The paired *t* test is obtained as follows:

```
> t.test(pre, post, paired=T)

        Paired t-test

data:   pre and post
t = 11.9414, df = 10, p-value = 3.059e-07
alternative hypothesis: true difference in means is not equal to 0
95 percent confidence interval:
 1074.072 1566.838
sample estimates:
mean of the differences
               1320.455
```

There is not much new to say about the output; it is virtually identical to that of a one-sample *t* test on the elementwise differences.

Notice that you have to specify `paired=T` explicitly in the call, indicating that you want a paired test. In the old-style interface for the unpaired *t* test, the two groups are specified as separate vectors and you would request that analysis by omitting `paired=T`. If data are actually paired, then it would be seriously inappropriate to analyze them without taking the pairing into account.

Even though it might be considered pedagogically dubious to show what you should *not* do, the following shows the results of an unpaired *t* test on the same data for comparison:

```
> t.test(pre, post) #WRONG!

        Welch Two Sample t-test

data:   pre and post
t = 2.6242, df = 19.92, p-value = 0.01629
alternative hypothesis: true difference in means is not equal to 0
```

```
95 percent confidence interval:
  270.5633 2370.3458
sample estimates:
mean of x mean of y
 6753.636   5433.182
```

The number symbol (or "hash") # introduces a comment in R. The rest of the line is skipped.

It is seen that t has become considerably smaller, although still significant at the 5% level. The confidence interval has become almost four times wider than in the correct paired analysis. Both illustrate the loss of efficiency caused by not using the information that the "pre" and "post" measurements are from the same persons. Alternatively, you could say that it demonstrates the gain in efficiency obtained by planning the experiment with two measurements on the same person, rather than having two independent groups of pre- and postmenstrual women.

4.7   The matched-pairs Wilcoxon test

The paired Wilcoxon test is the same as a one-sample Wilcoxon signed-rank test on the differences. The call is completely analogous to t.test:

```
> wilcox.test(pre, post, paired=T)
        Wilcoxon signed rank test with continuity correction

data:  pre and post
V = 66, p-value = 0.00384
alternative hypothesis: true mu is not equal to 0

Warning message:
Cannot compute exact p-value with ties in: wilcox.test.default(...
```

The result does not show any material difference from that of the t test. The p-value is not quite so extreme, not too surprising since the Wilcoxon rank sum cannot get any larger than it does when all differences have the same sign, whereas the t statistic can become arbitrarily extreme.

Again, we have trouble with tied data invalidating the exact p calculations. This time it is the two identical differences of -1540.

In the present case it is actually very easy to calculate the exact p-value for the Wilcoxon test. It is the probability of 11 positive differences + the probability of 11 negative ones: $2 \times (1/2)^{11} = 1/1024 = 0.00098$, so the approximate p-value is almost four times too large.

4.8 Exercises

4.1 Do the values of the `react` data set (notice that this is a single vector, not a data frame) look reasonably normally distributed? Does the mean differ significantly from zero, according to a t test?

4.2 In the data set `vitcap` use a t test to compare the vital capacity for the two groups. Calculate a 99% confidence interval for the difference. The result of this comparison may be misleading. Why?

4.3 Perform the analyses of the `react` and `vitcap` data using nonparametric techniques.

4.4 Perform the graphical checks of the assumptions for a paired t test in the `intake` data set.

4.5 The function `shapiro.test` computes a test of normality based on the degree of linearity of the Q–Q plot. Apply it to the `react` data. Does it help to remove the outliers?

4.6 The crossover trial in `ashina` can be analyzed for a drug effect in a simple way (how?) if you ignore a potential period effect. However, you can do better. Hint: Consider the intra-individual differences; if there were *only* a period effect present, how should the differences behave in the two groups? Compare the results of the simple method and the improved method.

4.7 Perform 10 one-sample t tests on simulated normally distributed data sets of 25 observations each. Repeat the experiment, but instead simulate samples from a different distribution; try the t distribution with 2 degrees of freedom and the exponential distribution (in the latter case test for mean equal to 1). Can you find a way to automate this so that you can have a larger number of replications?

5

Regression and correlation

The main object of this chapter is to show how to perform basic regression analyses, including plots for model checking and display of confidence and prediction intervals. Furthermore, we describe the related topic of correlation, in both its parametric and nonparametric variants.

5.1 Simple linear regression

We consider situations where you want to describe the relation between two variables using linear regression analysis. You may, for instance, be interested in describing `short.velocity` as a function of `blood.glucose`. This section deals only with the very basics, whereas several more complicated issues are postponed until Chapter 10.

The linear regression model is given by

$$y_i = \alpha + \beta x_i + \epsilon_i$$

in which the ϵ_i are assumed independent and $N(0, \sigma^2)$. The nonrandom part of the equation describes the y_i as lying on a straight line. The slope of the line (the *regression coefficient*) is β, the increase per unit change in x. The line intersects the y-axis at the *intercept* α.

The parameters α, β, and σ^2 can be estimated using the *method of least squares*: Find the values of α and β that minimize the sum of squared residuals

$$SS_{res} = \sum_i (y_i - (\alpha + \beta x_i))^2$$

This is not actually done by trial and error. It is quite simple to find closed-form expressions for the choice of parameters that gives the smallest value.

$$\hat{\beta} = \frac{\sum(x_i - \bar{x})(y_i - \bar{y})}{\sum(x_i - \bar{x})^2}$$
$$\hat{\alpha} = \bar{y} - \hat{\beta}\bar{x}$$

The residual variance is estimated as $SS_{res}/(n-2)$ and the residual standard deviation is of course the square root of that.

The empirical slope and intercept will deviate somewhat from the true values due to sampling variation. If you were to generate several sets of y_i at the same set of x_i, you would observe a distribution of empirical slopes and intercepts. Just like you could calculate the SEM to describe the variability of the empirical mean, it is also possible from a single sample of (x_i, y_i) to calculate the standard error of the computed estimates, s.e.$(\hat{\alpha})$ and s.e.$(\hat{\beta})$. These standard errors can be used to compute confidence intervals for the parameters and tests for whether a parameter has a specific value.

It is usually of prime interest to test the null hypothesis that $\beta = 0$ since that would imply that the line was horizontal and thus that the y have a distribution that is the same, whatever the value of x. You can compute a t test for that hypothesis simply by dividing the estimate by its standard error

$$t = \frac{\hat{\beta}}{s.e.(\hat{\beta})}$$

which follows a t distribution on $n - 2$ degrees of freedom if the true β is zero. A similar test can be calculated for whether the intercept is zero, but you should be aware that that is often a meaningless hypothesis, either because there is no natural reason to believe that the line should go through the origin, or because it would involve an extrapolation far outside the range of data.

For the example in this section, we need the data frame thuesen, which can be loaded and attached with

```
> data(thuesen)
> attach(thuesen)
```

For linear regression analysis, the function lm is used (*linear model*):

```
> lm(short.velocity~blood.glucose)

Call:
lm(formula = short.velocity ~ blood.glucose)

Coefficients:
  (Intercept)   blood.glucose
      1.09781         0.02196
```

The argument to lm is a *model formula*, in which the tilde symbol (~) should be read as "described by". This has been seen several times earlier, both in connection with boxplots and stripcharts and with the *t* and Wilcoxon tests.

The lm function handles much more complicated models than simple linear regression. There can be many other things besides a dependent and a descriptive variable in a model formula. A multiple linear regression analysis (which we discuss in Chapter 9) of, for example, y on x1, x2, and x3 is specified as y ~ x1 + x2 + x3.

In its raw form, the output of lm is very brief. All you see is the estimated intercept (α) and the estimated slope (β). The best-fitting straight line is seen to be short.velocity $= 1.098 + 0.0220 \times$ blood.glucose, but for instance no tests of significance are given.

The result of lm is a *model object*. This is a distinctive concept of the S language (of which R is a dialect). Where other statistical systems focus on generating printed output that can be controlled by setting options, you get instead the result of a model fit encapsulated in an object from which the desired quantities can be obtained using *extractor functions*. An lm object does in fact contain much more information than you see when it is printed.

A basic extractor function is summary:

```
> summary(lm(short.velocity~blood.glucose))

Call:
lm(formula = short.velocity ~ blood.glucose)

Residuals:
      Min        1Q    Median        3Q       Max
-0.40141  -0.14760  -0.02202   0.03001   0.43490

Coefficients:
              Estimate Std. Error t value Pr(>|t|)
(Intercept)    1.09781    0.11748   9.345 6.26e-09 ***
blood.glucose  0.02196    0.01045   2.101   0.0479 *
---
```

```
Signif. codes:  0 `***' 0.001 `**' 0.01 `*' 0.05 `.' 0.1 ` ' 1

Residual standard error: 0.2167 on 21 degrees of freedom
Multiple R-Squared: 0.1737,      Adjusted R-squared: 0.1343
F-statistic: 4.414 on 1 and 21 DF, p-value: 0.0479
```

The above is a format that looks more like what other statistical packages would output. The following is a "dissection" of the output:

```
Call:
lm(formula = short.velocity ~ blood.glucose)
```

Like in t.test, etc., the output starts with something that is essentially a repeat of the function call. This is not very interesting when one has just given it as a command to R, but it is useful if the result is saved in a variable that is printed later.

```
Residuals:
     Min       1Q    Median       3Q       Max
-0.40141 -0.14760 -0.02202  0.03001  0.43490
```

This gives a superficial view of the distribution of the residuals, which may be used as a quick check of the distributional assumptions. The average of the residuals is zero by definition, so the median should not be far from zero and the minimum and maximum should roughly be equal in absolute value. In the example it can be noticed that the third quartile is remarkably close to zero, but in view of the small number of observations, this is not really something to worry about.

```
Coefficients:
              Estimate Std. Error t value Pr(>|t|)
(Intercept)    1.09781    0.11748   9.345 6.26e-09 ***
blood.glucose  0.02196    0.01045   2.101   0.0479 *
---
Signif. codes:  0 `***' 0.001 `**' 0.01 `*' 0.05 `.' 0.1 ` ' 1
```

Here we see the regression coefficient and the intercept again, but this time with accompanying standard errors, t tests, and p-values. The symbols to the right are graphical indicators of the level of significance. The line below the table shows the definition of these indicators; one star means $0.01 < p < 0.05$.

The graphical indicators have been the target of some controversy. Some people like to have the possibility of seeing at a glance whether there is "anything interesting" in an analysis, whereas others feel that the indicators too often correspond to meaningless tests. For instance, the intercept in the above analysis is hardly a meaningful quantity at all, and the three-star significance of it is certainly irrelevant. If you are bothered by the stars, turn them off with options(show.signif.stars=FALSE).

```
Residual standard error: 0.2167 on 21 degrees of freedom
```

This is the residual variation, an expression of the variation of the observations around the regression line, estimating the model parameter σ.

```
Multiple R-Squared: 0.1737,  Adjusted R-squared: 0.1343
```

The first item above is R^2, which in a simple linear regression may be recognized as the squared Pearson correlation coefficient (see Section 5.4.1), that is, $R^2 = r^2$. The other one is the adjusted R^2; if you multiply it by 100%, it can be interpreted as "% variance reduction" (this can, in fact, become negative).

```
F-statistic: 4.414 on 1 and 21 DF, p-value: 0.0479
```

This is an F test for the hypothesis that the regression coefficient is zero. This test is not really interesting in a simple linear regression analysis since it just duplicates information already given — it becomes more interesting when there is more than one explanatory variable. Notice that it gives the exact same result as the t test for a zero slope. In fact, the F test is identical to the square of the t test: $4.414 = (2.101)^2$. This is true in any model with 1 degree of freedom.

We see ahead how to draw residual plots and plots of data with confidence and prediction limits. First, we draw just the points and the fitted line. Figure 5.1 has been constructed as follows:

```
> plot(blood.glucose,short.velocity)
> abline(lm(short.velocity~blood.glucose))
```

abline draws lines based on the intercept and slope (a and b, hence the name). It can be used with scalar values as in abline(1.1,0.022), but conveniently it can also extract the information from a linear model fitted to data with lm.

5.2 Residuals and fitted values

We have seen how summary can be used to extract information about the results of a regression analysis. Two further extraction functions are fitted and resid. They are used as follows. For convenience, we first store the value returned by lm under the name lm.velo (short for "velocity", but you could of course use any other name).

```
> lm.velo <- lm(short.velocity~blood.glucose)
```

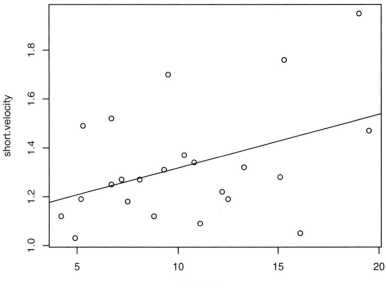

Figure 5.1. Scatterplot with regression line.

```
> fitted(lm.velo)
       1        2        3        4        5        6        7
1.433841 1.335010 1.275711 1.526084 1.255945 1.214216 1.302066
       8        9       10       11       12       13       14
1.341599 1.262534 1.365758 1.244964 1.212020 1.515103 1.429449
      15       17       18       19       20       21       22
1.244964 1.190057 1.324029 1.372346 1.451411 1.389916 1.205431
      23       24
1.291085 1.306459
> resid(lm.velo)
           1            2            3            4            5
 0.326158532  0.004989882 -0.005711308 -0.056084062  0.014054962
           6            7            8            9           10
 0.275783754  0.007933665 -0.251598875 -0.082533795 -0.145757649
          11           12           13           14           15
 0.005036223 -0.022019994  0.434897199 -0.149448964  0.275036223
          17           18           19           20           21
-0.070057471  0.045971143 -0.182346406 -0.401411486 -0.069916424
          22           23           24
-0.175431237 -0.171085074  0.393541161
```

The function fitted returns fitted values — the *y*-values that you
would expect for the given *x*-values according to the best-fitting straight
line: in the present case 1.098+0.0220*blood.glucose. The resid-

uals shown by resid is the difference between this and the observed
short.velocity.

Note that the fitted values and residuals are labeled with the row names
of the thuesen data frame. Notice in particular that they do not contain
observation no. 16, which had a missing value in the response variable.

It is necessary to discuss some awkward aspects that arise when there are
missing values in data.

To put the fitted line on the plot, you might, although it is easier to use
abline(lm.velo), get the idea of doing it with lines, *but*

```
> plot(blood.glucose,short.velocity)
> lines(blood.glucose,fitted(lm.velo))
Error in xy.coords(x, y) : x and y lengths differ
```

— which is true. There are 24 observations but only 23 fitted values because one of the short.velocity values is NA. So we would need
blood.glucose, but only for those patients whose short.velocity
has been recorded.

```
> lines(blood.glucose[!is.na(short.velocity)],fitted(lm.velo))
```

Recall that the is.na function yields a vector that is TRUE wherever the
argument is NA (missing). One advantage to this method is that the fitted
line does not extend beyond the range of data. The technique works but
becomes clumsy if there are missing values in several variables:

```
...blood.glucose[!is.na(short.velocity) & !is.na(blood.glucose)]...
```

It becomes easier with the function complete.cases, which can find
observations that are nonmissing on several variables or across an entire
data frame.

```
> cc <- complete.cases(thuesen)
```

We could then attach thuesen[cc,] and work on from there. However,
there is a better alternative available in using the na.exclude method for
NA handling. This can be set either as an argument to lm or as an option,
that is,

```
> options(na.action=na.exclude)
> lm.velo <- lm(short.velocity~blood.glucose)
> fitted(lm.velo)
       1        2        3        4        5        6        7
1.433841 1.335010 1.275711 1.526084 1.255945 1.214216 1.302066
       8        9       10       11       12       13       14
1.341599 1.262534 1.365758 1.244964 1.212020 1.515103 1.429449
```

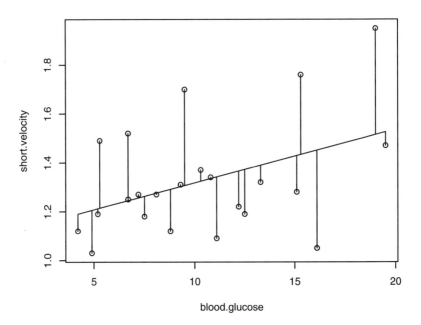

Figure 5.2. Scatterplot of short.velocity versus blood.glucose with fitted line and residual line segments.

```
       15         16         17         18         19         20         21
1.244964         NA   1.190057   1.324029   1.372346   1.451411   1.389916
       22         23         24
1.205431   1.291085   1.306459
```

Notice how the missing observation, no. 16, now appears in the fitted values with a missing fitted value. It is necessary to recalculate the lm.velo object after changing the option.

To create a plot where residuals are displayed by connecting observations to corresponding points on the fitted line, you can do the following. The final result will look like Figure 5.2. segments draws line segments; its arguments are the endpoint coordinates in the order (x_1, y_1, x_2, y_2).

```
> segments(blood.glucose,fitted(lm.velo),
+          blood.glucose,short.velocity)
```

A simple plot of residuals versus fitted values is obtained like this (Figure 5.3):

```
> plot(fitted(lm.velo),resid(lm.velo))
```

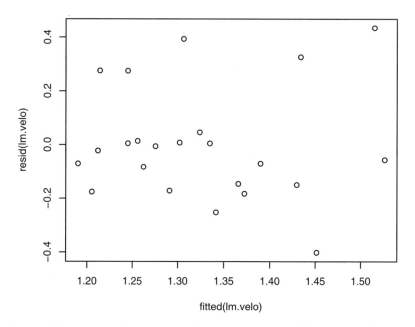

Figure 5.3. `short.velocity` and `blood.glucose`: residuals versus fitted value.

and we can get an indication of whether residuals might have come from a normal distribution by checking for a straight line on a Q–Q plot (see Section 3.2.3) as follows (Figure 5.4):

```
> qqnorm(resid(lm.velo))
```

5.3 Prediction and confidence bands

Fitted lines are often presented with uncertainty bands around them. There are two kinds of bands, often referred to as the "narrow" and "wide" limits.

The narrow bands, *confidence bands*, reflect the uncertainty about the line itself, like the SEM expresses the precision with which a mean is known. If there are many observations, the bands will be quite narrow, reflecting a well-determined line. These bands often show a marked curvature since the line is better determined near the center of the point cloud. This is a fact that can be shown mathematically, but you may also understand it intuitively as follows: The predicted value at $\bar{x}$ will be $\bar{y}$, whatever the

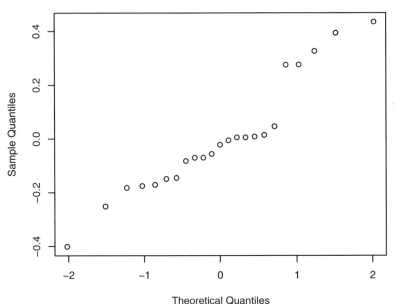

Figure 5.4. `short.velocity` and `blood.glucose`: Q–Q plot of residuals.

slope is, and hence the standard error of the fitted value at that point is the SEM of the ys. At other values of x there will also be a contribution from the variability of the estimated slope, having increasing influence as you move away from $\bar{x}$. Technically, you also need to establish that $\bar{y}$ and $\hat{\beta}$ are uncorrelated.

The wide bands, *prediction bands*, include the uncertainty about future observations. These bands should capture the majority of the observed points and will not collapse to a line as the number of observations increases. Rather, the limits approach the true line ± 2 standard deviations (for 95% limits). In smaller samples, the bands do curve since they include uncertainty about the line itself, but not as markedly as the confidence bands. Obviously, these limits rely strongly on the assumption of normally distributed errors with a constant variance, so you should not use such limits unless you believe that it is a reasonable approximation for the data at hand.

Predicted values, with or without prediction and confidence bands, may be extracted with the function `predict`. With no arguments, it just gives the fitted values:

```
> predict(lm.velo)
```

1	2	3	4	5	6	7
1.433841	1.335010	1.275711	1.526084	1.255945	1.214216	1.302066
8	9	10	11	12	13	14
1.341599	1.262534	1.365758	1.244964	1.212020	1.515103	1.429449
15	16	17	18	19	20	21
1.244964	NA	1.190057	1.324029	1.372346	1.451411	1.389916
22	23	24				
1.205431	1.291085	1.306459				

If you add interval="confidence" or interval="prediction"
then you get the vector of predicted values augmented with limits. The
arguments can be abbreviated:

```
> predict(lm.velo,int="c")
        fit      lwr      upr
1  1.433841 1.291371 1.576312
2  1.335010 1.240589 1.429431
...
23 1.291085 1.191084 1.391086
24 1.306459 1.210592 1.402326
> predict(lm.velo,int="p")
        fit       lwr      upr
1  1.433841 0.9612137 1.906469
2  1.335010 0.8745815 1.795439
...
23 1.291085 0.8294798 1.752690
24 1.306459 0.8457315 1.767186
```

fit is the expected values, here identical to the fitted values (it need not
be; read on). lwr and upr (*lower/upper*) are the confidence limits for the
expected values, respectively, the prediction limits for short.velocity
for new persons with these values of blood.glucose.

The best way to add such intervals to a scatterplot is to use the matlines
function, which plots the columns of a matrix against a vector.

There are a few snags to this, however: (a) The blood.glucose values
are in random order; we do not want line segments connecting points
haphazardly along the confidence curves; (b) the prediction limits, particularly the lower one, extend outside the plot region; and (c) the matlines
command needs to be prevented from cycling through line styles and
colours. Notice that the na.exclude setting (p. 101) prevents us from
also having an observation omitted from the predicted values.

The solution is to *predict in a new data frame*, containing suitable *x* values
(here blood.glucose) at which to predict. It is done as follows:

```
> pred.frame <- data.frame(blood.glucose=4:20)
> pp <- predict(lm.velo, int="p", newdata=pred.frame)
> pc <- predict(lm.velo, int="c", newdata=pred.frame)
> plot(blood.glucose,short.velocity,
```

```
+        ylim=range(short.velocity, pp, na.rm=T))
> pred.gluc <- pred.frame$blood.glucose
> matlines(pred.gluc, pc, lty=c(1,2,2), col="black")
> matlines(pred.gluc, pp, lty=c(1,3,3), col="black")
```

This is what happens: First we create a new data frame in which the blood.glucose variable contains the values at which we want predictions to be made. pp and pc are then made to contain the result of predict for the new data in pred.frame with prediction limits and confidence limits, respectively.

Now for the plotting: First we create a standard scatterplot, except that we ensure that it has enough room for the prediction limits. This is obtained by setting ylim=range(short.velocity, pp, na.rm=T). The function range returns a vector of length 2, containing the minimum and maximum value of its arguments. We need the na.rm=T argument to cause missing values to be skipped for the range computation; notice that short.velocity is included to ensure that points outside the prediction limits are not missed (although in this case there are none). Finally, the curves are added, using as x-values the blood.glucose used for the prediction, and setting the line types and colours to more sensible values. The final result is seen in Figure 5.5.

5.4   Correlation

A correlation coefficient is a symmetric, scale-invariant measure of association between two random variables. It ranges from -1 to $+1$, where the extremes indicate perfect correlation and 0 means no correlation. The sign is negative when large values of one variable are associated with small values of the other and positive if both variables tend to be large or small simultaneously. The reader should be warned that there are many incorrect uses of correlation coefficients, particularly when they are used in regression-type settings.

This section describes the computation of parametric and nonparametric correlation measures in R.

5.4.1   Pearson correlation

The Pearson correlation is rooted in the two-dimensional normal distribution where the theoretical correlation describes the contour ellipses for the density. If both variables are scaled to have a variance of 1, then a correlation of zero corresponds to circular contours, whereas the ellipses

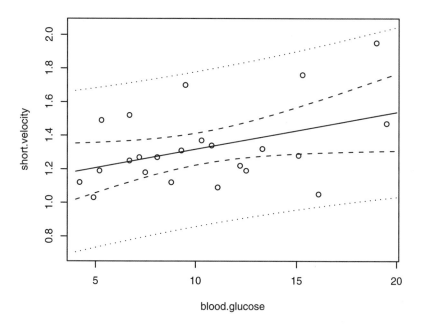

Figure 5.5. Plot with confidence and tolerance bands.

become narrower and finally collapse into a line segment as the correlation approaches ± 1.

The empirical correlation coefficient is

$$r = \frac{\sum(x_i - \bar{x})(y_i - \bar{y})}{\sqrt{\sum(x_i - \bar{x})^2 \sum(y_i - \bar{y})^2}}$$

It can be shown that $|r|$ will be less than 1 unless there is a perfect linear relation between x_i and y_i, and for that reason the Pearson correlation is sometimes called the "linear correlation".

It is possible to test the significance of the correlation by transforming it to a t-distributed variable (the formula is not particularly elucidating so we skip it here), which will be identical with the test obtained from testing significance of the slope of either the regression of y on x, or vice versa.

The function cor can be used to compute the correlation between two or more vectors. However, if it is naively applied to the two vectors in thuesen, the following happens:

```
> cor(blood.glucose,short.velocity)
Error in cor(blood.glucose, short.velocity) :
```

```
missing observations in cov/cor
```

All the elementary statistical functions in R require either that all values are nonmissing or that you explicitly state what should be done with the cases with missing values. For mean, var, sd, and similar one-vector functions, you can give the argument na.rm=T to indicate that missing values should be removed before the computation. For cor you can write

```
> cor(blood.glucose,short.velocity,use="complete.obs")
[1] 0.4167546
```

The reason that cor does not use na.rm=T like the other functions is that there are more possibilities than just removing incomplete cases or failing. If more than two variables are in play, it is also possible to use information from all nonmissing *pairs* of measurements (this might result in a correlation matrix that is not positive definite, though).

You can obtain the entire matrix of correlations between all variables in a data frame by saying, for instance

```
> cor(thuesen,use="complete.obs")
                blood.glucose short.velocity
blood.glucose      1.0000000      0.4167546
short.velocity     0.4167546      1.0000000
```

Of course, this is more interesting when the data frame contains more than two vectors!

However, the above calculations give no indication of whether the correlation is significantly different from zero. To that end, you need cor.test. It works simply by specifying the two variables:

```
> cor.test(blood.glucose,short.velocity)

        Pearson's product-moment correlation

data:  blood.glucose and short.velocity
t = 2.101, df = 21, p-value = 0.0479
alternative hypothesis: true correlation is not equal to 0
sample estimates:
      cor
0.4167546
```

Notice that it is exactly the same p-value as in the regression analysis in Section 5.1 and also as that based on the ANOVA table for the regression model, which is described in Section 6.5.

5.4.2 Spearman's ρ

As with the one- and two-sample problems, you may be interested in nonparametric variants. These have the advantage of not depending on the normal distribution and, indeed, being invariant to monotone transformations of the coordinates. The main disadvantage is that its interpretation is not quite clear. A popular and simple choice is Spearman's rank correlation coefficient ρ. This is obtained quite simply by replacing the observations by their rank and computing the correlation. Under the null hypothesis of independence between the two variables the exact distribution of ρ can be calculated.

Unlike group comparisons where there is essentially one function per named test, correlation tests are all grouped into cor.test. There is no special spearman.test function. Instead, the test is considered one of several possibilities for testing correlations and is therefore specified via an option to cor.test:

```
> cor.test(blood.glucose,short.velocity,method="spearman")

        Spearman's rank correlation rho

data:  blood.glucose and short.velocity
S = 1380, p-value = 0.139
alternative hypothesis: true rho is not equal to 0
sample estimates:
     rho
0.318002

Warning message:
p-values may be incorrect due to ties in: cor.test.default(...
```

5.4.3 Kendall's τ

The third correlation method that you can choose is Kendall's τ, which is based on counting the number of *concordant* and *discordant* pairs. A pair of points is concordant if the difference in the x-coordinate is of the same sign as the difference in the y-coordinate. For a perfect monotone relation, either all pairs will be concordant or all pairs will be discordant. Under independence, there should be as many concordant pairs as there are discordant.

Since there are many pairs of points to check, this is quite a computationally intensive procedure, compared to the two others. In small data sets like the present, it does not matter at all, though, and it is generally usable up to at least 5000 observations.

The τ coefficient has the advantage of a more direct interpretation over Spearman's ρ, but apart from that there is little reason to prefer one over the other.

```
> cor.test(blood.glucose,short.velocity,method="kendall")

        Kendall's rank correlation tau

data:   blood.glucose and short.velocity
z.tau = 1.5706, p-value = 0.1163
alternative hypothesis: true tau is not equal to 0
sample estimates:
      tau
0.2350616

Warning message:
Cannot compute exact p-value with ties in: cor.test.default(...
```

Notice that neither of the two nonparametric correlations is significant at the 5% level, which the Pearson correlation is, albeit only borderline significant.

5.5   Exercises

5.1   With the rmr data set, plot metabolic rate versus body weight. Fit a linear regression model to the relation. According to the fitted model, what is the predicted metabolic rate for a body weight of 70 kg? Give a 95% confidence interval for the slope of the line.

5.2   In the juul data set fit a linear regression model to the square root of the IGF-I concentration versus age, to the group of subjects over 25 years old.

5.3   In the malaria data set analyze the log-transformed antibody level versus age. Make a plot of the relation. Do you notice anything peculiar?

5.4   One can generate simulated data from the two-dimensional normal distribution as follows: (a) Generate X as a normal variate with mean 0 and standard deviation 1; (b) generate Y with mean ρX and standard deviation $\sqrt{1 - \rho^2}$. Use this to create scatterplots of simulated data with a given correlation. Compute the Spearman and Kendall statistics for some of these data sets.

6

Analysis of variance and the Kruskal–Wallis test

In this section we consider comparisons among more than two groups parametrically, using analysis of variance, as well as nonparametrically, using the Kruskal–Wallis test. Furthermore, we see two-way analysis of variance in the case of one observation per cell.

6.1 One-way analysis of variance

We start this section with a brief sketch of the theory underlying the one-way analysis of variance. A little bit of notation is necessary. Let x_{ij} denote observation no. j in group i so that x_{35} is the fifth observation in group 3; $\bar{x}_i$ is the mean for group i and $\bar{x}.$ is the grand mean (average of all observations).

We can decompose the observations as

$$x_{ij} = \bar{x}. + \underbrace{(\bar{x}_i - \bar{x}.)}_{\substack{\text{deviation of} \\ \text{group mean from} \\ \text{grand mean}}} + \underbrace{(x_{ij} - \bar{x}_i)}_{\substack{\text{deviation of} \\ \text{observation from} \\ \text{group mean}}}$$

informally corresponding to the model

$$X_{ij} = \mu + \alpha_i + \epsilon_{ij}, \qquad \epsilon_{ij} \sim N(0, \sigma^2)$$

in which the hypothesis that all the groups are the same implies that all α_i are zero. Notice that the error terms ϵ_{ij} are assumed to be independent and have the same variance.

Now consider the sums of squares of the underbraced terms, known as *variation within groups*

$$\text{SSD}_W = \sum_i \sum_j (x_{ij} - \bar{x}_i)^2$$

and *variation between groups*

$$\text{SSD}_B = \sum_i \sum_j (\bar{x}_i - \bar{x}_.)^2 = \sum_i n_i (\bar{x}_i - \bar{x}_.)^2$$

It is possible to prove that

$$\text{SSD}_B + \text{SSD}_W = \text{SSD}_{\text{total}} = \sum_i \sum_j (x_{ij} - \bar{x}_.)^2$$

That is, the total variation is split into a term describing differences between group means and a term describing differences between individual measurements within the groups. One says that the grouping explains part of the total variation, and obviously an informative grouping will explain a large part of the variation.

However, the sums of squares can only be positive, so even a completely irrelevant grouping will always "explain" some part of the variation. The question is how small an amount of explained variation can be before it might as well be due to chance. It turns out that in the absence of any systematic differences between the groups, you should expect the sum of squares to be partitioned according to the degrees of freedom for each term: $k - 1$ for SSD_B and $N - k$ for SSD_W, where k is the number of groups and N is the total number of observations.

Accordingly, you can normalize the sums of squares, by calculating *mean squares*:

$$\text{MS}_W = \text{SSD}_W/(N - k)$$
$$\text{MS}_B = \text{SSD}_B/(k - 1)$$

MS_W is the pooled variance obtained by combining the individual group variances and thus an estimate of σ^2. In the absence of a true group effect, MS_B will also be an estimate of σ^2, but if there *is* a group effect, then the differences between group means and hence MS_B will tend to be larger. Thus, a test for significant differences between the group means can be performed by comparing two variance estimates. This is why the procedure is called *analysis of variance* even though the objective is to compare the group means.

A formal test needs to account for the fact that random variation will cause some difference in the mean squares. You calculate

$$F = MS_B/MS_W$$

so that F is ideally 1, but some variation around that value is expected. The distribution of F under the null hypothesis is an F distribution with $k - 1$ and $N - k$ degrees of freedom. You reject the hypothesis of identical means if F is larger than the 95% quantile in that F distribution (if the significance level is 5%). Notice that this test is one-sided; a very small F would occur if the group means were very similar and that will of course not signify a difference between the groups.

Simple analyses of variance can be performed in R using the function lm, which is also used for regression analysis. For more elaborate analyses, there are also the functions aov and lme (linear mixed effects models, from the nlme package). An implementation of Welch's procedure, relaxing the assumption of equal variances and generalizing the unequal-variance t test, is implemented in oneway.test (see Section 6.1.2).

The main example in this section is the "red cell folate" data from Altman (1991, p. 208). To use lm it is necessary to have the data values in one vector and a factor variable (see Section 1.2.7) describing the division into groups. The red.cell.folate data set contains a data frame in the proper format.

```
> data(red.cell.folate)
> attach(red.cell.folate)
> summary(red.cell.folate)
      folate            ventilation
 Min.   :206.0    N20+02,24h:8
 1st Qu.:249.5    N20+02,op :9
 Median :274.0    02,24h    :5
 Mean   :283.2
 3rd Qu.:305.5
 Max.   :392.0
```

Recall that summary applied to a data frame gives a short summary of the distribution of each of the variables contained in it. The format of the summary is different for numeric vectors and factors, so that provides a check that the variables are defined correctly.

The category names for ventilation mean "N_2O and O_2 for 24 hours", "N_2O and O_2 during operation", and "only O_2 for 24 hours".

In the following the analysis of variance is demonstrated first, and then a couple of useful techniques for the presentation of grouped data as tables and graphs are shown.

The specification of a one-way analysis of variance is analogous to a regression analysis. The only difference is that the descriptive variable needs to be a factor and not a numeric variable. We calculate a model object using lm and extract the analysis of variance table with anova.

```
> anova(lm(folate~ventilation))
Analysis of Variance Table

Response: folate
            Df Sum Sq Mean Sq F value  Pr(>F)
ventilation  2  15516    7758  3.7113 0.04359 *
Residuals   19  39716    2090
---
Signif. codes:  0 '***' 0.001 '**' 0.01 '*' 0.05 '.' 0.1 ' ' 1
```

Here we have SSD_B and MS_B in the top line and SSD_W and MS_W in the second line.

In statistical textbooks the sums of squares are most often labeled "between groups" and "within groups". Like most other statistical software, R uses a slightly different labeling. Variation between groups is labeled by the name of the grouping factor (ventilation) and variation within groups is labelled Residual. ANOVA tables can be used for a wide range of statistical models, and it is convenient to use a format that is less linked to the particular problem of comparing groups.

For a further example, consider the data set juul, introduced in Section 3.1. Notice that the tanner variable in this data set is a numeric vector and not a factor. For purposes of tabulation this makes little difference, but it would be a serious error to use it in this form in an analysis of variance:

```
> data(juul)
> attach(juul)
> anova(lm(igf1~tanner))   ## WRONG!
Analysis of Variance Table

Response: igf1
           Df   Sum Sq  Mean Sq F value    Pr(>F)
tanner      1 10985605 10985605  686.07 < 2.2e-16 ***
Residuals 790 12649728    16012
---
Signif. codes:  0 '***' 0.001 '**' 0.01 '*' 0.05 '.' 0.1 ' ' 1
```

This does not describe a grouping of data but a linear regression on the group number! Notice the telltale 1 DF for the effect of tanner.

Things can be fixed as follows:

```
> juul$tanner <- factor(juul$tanner,
```

```
+                        labels=c("I","II","III","IV","V"))
> detach(juul)
> attach(juul)
> summary(tanner)
   I   II  III   IV    V NA's
 515  103   72   81  328  240
> anova(lm(igf1~tanner))
Analysis of Variance Table

Response: igf1
           Df   Sum Sq  Mean Sq F value    Pr(>F)
tanner      4 12696217  3174054  228.35 < 2.2e-16 ***
Residuals 787 10939116    13900
---
Signif. codes:  0 '***' 0.001 '**' 0.01 '*' 0.05 '.' 0.1 ' ' 1
```

We needed to reattach the `juul` data frame in order to use the changed
definition. An attached data frame is effectively a separate copy of it
(although it does not take up extra space as long as the original is un-
changed). The `Df` column has an entry of 4 for `tanner` now, as it should
have.

6.1.1 Pairwise comparisons and multiple testing

If the F test shows that there is a difference between groups, the question quickly arises of wherein the difference lies. It becomes necessary to
compare the individual groups.

Part of this information can be found in the regression coefficients. You can
use `summary` to extract regression coefficients with standard errors and t
tests. These coefficients do not have their usual meaning as the slope of a
regression line but have a special interpretation, which is described below.

```
> summary(lm(folate~ventilation))

Call:
lm(formula = folate ~ ventilation)

Residuals:
    Min     1Q  Median     3Q    Max
-73.625 -35.361  -4.444 35.625 75.375

Coefficients:
                    Estimate Std. Error t value Pr(>|t|)
(Intercept)           316.63      16.16  19.588 4.65e-14 ***
ventilationN2O+O2,op  -60.18      22.22  -2.709   0.0139 *
ventilationO2,24h     -38.63      26.06  -1.482   0.1548
---
```

```
Signif. codes:  0 '***' 0.001 '**' 0.01 '*' 0.05 '.' 0.1 ' ' 1

Residual standard error: 45.72 on 19 degrees of freedom
Multiple R-Squared: 0.2809,    Adjusted R-squared: 0.2052
F-statistic: 3.711 on 2 and 19 DF,  p-value: 0.04359
```

The interpretation of the estimates is that the intercept is the mean in the first group (N2O+O2, 24h), whereas the other two describe the *difference* between the relevant group and the first one.

There are multiple ways of representing the effect of a factor variable in linear models (and one-way analysis of variance is the simplest example of a linear model with a factor variable). The representations are in terms of *contrasts*, the choice of which can be controlled either by global options or as part of the model formula. We do not go deeply into this; just mention that the contrasts used by default are the so-called *treatment contrasts* in which the first group is treated as a baseline and the other groups are given relative to that. Concretely, the analysis is performed as a multiple regression analysis (see Chapter 9) by introducing two *dummy variables*, which are 1 for observations in the relevant group and 0 elsewhere.

Among the *t* tests in the table, you can immediately find a test for the hypothesis that the first two groups have the same true mean ($p = 0.0139$) and also whether the first and the third might be identical ($p = 0.1548$). However, a comparison of the last two groups cannot be found. This can be overcome by modifying the factor definition (see the help page for `relevel`), but that gets tedious when there are more than a few groups.

If we want to compare all groups, we ought to correct for *multiple testing*. Performing many tests will increase the probability of finding one of them to be significant, that is, the *p*-values tend to be exaggerated. A common adjustment method is the *Bonferroni correction*, which is based on the fact that the probability of observing at least one of n events is less than the sum of the probabilities for each event. Thus, by dividing the significance level by the number of tests or, equivalently, multiplying the *p*-values, we obtain a *conservative* test where the probability of a significant result is less than or equal to the formal significance level.

A function called `pairwise.t.test` computes all possible two-group comparisons. It is also capable of making adjustments for multiple comparisons and works like this:

```
> pairwise.t.test(folate, ventilation, p.adj="bonferroni")

        Pairwise comparisons using t tests with pooled SD

data:   folate and ventilation
```

```
          N20+02,24h N20+02,op
N20+02,op 0.042       -
02,24h    0.464       1.000
```

```
P value adjustment method: bonferroni
```

The output is a table of p-values for the pairwise comparisons. Here, the p-values have been adjusted by the Bonferroni method, where the unadjusted values have been multiplied by the number of comparisons, namely 3. If that results in a value bigger than 1, then the adjustment procedure sets the adjusted p-value to 1.

The default method for `pairwise.t.test` is actually not the Bonferroni correction but a variant due to Holm. In this method only the smallest p needs to be corrected by the full number of tests, the second smallest is corrected by $n - 1$, etc. — unless that would make it smaller than the previous one since the order of the p-values should be unaffected by the adjustment.

```
> pairwise.t.test(folate,ventilation)

        Pairwise comparisons using t tests with pooled SD

data:   folate and ventilation

          N20+02,24h N20+02,op
N20+02,op 0.042       -
02,24h    0.310       0.408

P value adjustment method: holm
```

6.1.2 Relaxing the variance assumption

The traditional one-way ANOVA requires an assumption of equal variances for all groups. There is, however, an alternative procedure that does not require that assumption. It is due to Welch and similar to the unequal variances t test. This has been implemented in the `oneway.test` function:

```
> oneway.test(folate~ventilation)

        One-way analysis of means (not assuming equal variances)

data:   folate and ventilation
F = 2.9704, num df =  2.000, denom df = 11.065, p-value = 0.09277
```

In this case the *p*-value increased to a nonsignificant value, presumably related to the fact that the group that seems to differ from the other two also has the largest variance.

It is also possible to perform the pairwise *t* tests so that they do not use a common pooled standard deviation. This is controlled by the argument `pool.sd`.

```
>  pairwise.t.test(folate,ventilation,pool.sd=F)

         Pairwise comparisons using t tests with non-pooled SD

data:   folate and ventilation

            N20+02,24h N20+02,op
N20+02,op 0.087         -
02,24h    0.321      0.321

P value adjustment method: holm
```

Again, it is seen that the significance disappears as we remove the constraint on the variances.

6.1.3   Graphical presentation

Of course, there are many ways to present grouped data. Here we create a somewhat elaborate plot where the raw data are plotted as a stripchart and overlaid with an indication of means and SEMs (Figure 6.1):

```
> xbar <- tapply(folate, ventilation, mean)
> s <- tapply(folate, ventilation, sd)
> n <- tapply(folate, ventilation, length)
> sem <- s/sqrt(n)
> stripchart(folate~ventilation,"jitter",jit=0.05,pch=16,vert=T)
> arrows(1:3,xbar+sem,1:3,xbar-sem,angle=90,code=3,length=.1)
> lines(1:3,xbar,pch=4,type="b",cex=2)
```

Here we used pch=16 (small plotting dots) in `stripchart` and put vertical=T to make the "strips" vertical.

The error bars have been made with `arrows`, which adds arrows to a plot. We slightly abuse the fact that the angle of the arrowhead is adjustable to create the little crossbars at either end. The first four arguments specify the endpoints, (x_1, y_1, x_2, y_2); the angle argument gives the angle between the lines of the arrowhead and -shaft, here set to 90°; and length is the length of the arrowhead (in inches on a printout). Finally, code=3 means that the arrow should have a head at both ends. Note that the *x* coordinates of the stripcharts are simply the group numbers.

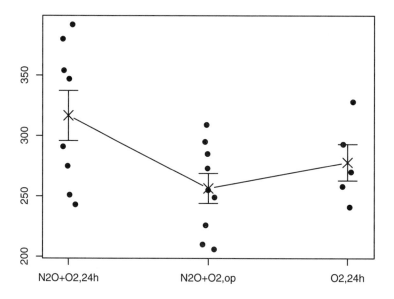

Figure 6.1. "Red cell folate" data with $\bar{x} \pm 1$ SEM.

The indication of averages and the connecting lines are done with `lines`, where `type="b"` (*both*) means that both points and lines are printed, leaving gaps in the lines to make room for the symbols. `pch=4` is a cross and `cex=2` requests that the symbols be drawn in double size.

It is debatable whether you should draw the plot using 1 SEM as is done here or whether perhaps it is better to draw proper confidence intervals for the means (approximately 2 SEM), or maybe even SD instead of SEM. The latter point has to do with whether the plot is to be used in a descriptive or an analytical manner. SEMs are not useful for describing the distributions in the groups; they only say how precisely the mean is determined. On the other hand, SDs do not enable the reader to see at a glance which groups are significantly different.

In many fields it appears to have become the tradition to use 1 SEM "because they are the smallest", that is, it makes differences look more dramatic. Probably, the best thing to do is to follow the traditions in the relevant field and "calibrate your eyeballs" accordingly.

One word of warning, though: At small group sizes, the rule of thumb that the confidence interval is mean $\pm$ 2 SEM becomes badly misleading. At a group size of 2, it actually has to be 12.7 SEM! and that is a correction depending heavily on data having the normal distribution. If you have such small groups, it may be advisable to use a pooled SD for the entire data set rather than the group-specific SDs. This does, of course, require that you can reasonably assume that the true standard deviation actually is the same in all groups.

6.1.4   Bartlett's test

Testing whether the distribution of a variable has the same variance in all groups can be done using Bartlett's test, although like the F test for comparison of two variances, it is rather nonrobust against departures from the assumption of normal distributions. As in `var.test` it is assumed that the data are from independent groups. The procedure is performed as follows:

```
> bartlett.test(folate~ventilation)

        Bartlett test for homogeneity of variances

data:   folate by ventilation
Bartlett's K-squared = 2.0951, df = 2, p-value = 0.3508
```

— that is, in this case, nothing in data contradicts the assumption of equal variances in the three groups.

6.2   Kruskal–Wallis test

A nonparametric counterpart of a one-way analysis of variance is the Kruskal–Wallis test. As in the Wilcoxon two-sample test (see Section 4.5), data are replaced with their ranks without regard to the grouping, only this time the test is based on the between-group sum of squares calculated from the average ranks. Again, the distribution of the test statistic can be worked out based on the idea that under the hypothesis of irrelevant grouping, the problem reduces to a combinatorial one of sampling the within-group ranks from a fixed set of numbers.

You can make R calculate the Kruskal–Wallis test as follows:

```
> kruskal.test(folate~ventilation)

        Kruskal-Wallis rank sum test

data:   folate by ventilation
Kruskal-Wallis chi-squared = 4.1852, df = 2, p-value = 0.1234
```

It is seen that there is no significant difference using this test. This should not be too surprising in view of the fact that the F test in the one-way analysis of variance was only borderline significant. Also, the Kruskal–Wallis test is less efficient than its parametric counterpart if the assumptions hold, although it does not invariably give a larger p-value.

6.3   Two-way analysis of variance

One-way analysis of variance deals with one-way classifications of data. It is also possible to analyze data that are cross-classified according to several criteria. When a cross-classified design is *balanced*, then you can almost read the entire statistical analysis from a single analysis of variance table, and that table generally consists of items that are simple to compute — something that was very important before the computer era. Balancedness is a concept that is hard to define exactly; for a two-way classification, a sufficient condition is that the cell counts are equal, but there are other balanced designs.

Here we restrict ourselves to the case of a single observation per cell. This typically arises from having multiple measurements on the same experimental unit and in this sense generalizes the paired t test.

Let x_{ij} denote the observation in row i and column j of the $m \times n$ table. This is similar to the notation used for one-way analysis of variance, but notice that there is now a connection between observations with the same j, so that it makes sense to look at both row averages $\bar{x}_{i.}$ and column averages $\bar{x}_{.j}$. Consequently, it now makes sense to look at both *variation between rows*:

$$\mathrm{SSD}_R = n \sum_i (\bar{x}_{i.} - \bar{x}_{..})^2$$

and *variation between columns*:

$$\mathrm{SSD}_C = m \sum_j (\bar{x}_{.j} - \bar{x}_{..})^2$$

Subtracting these two from the total variation leaves the *residual variation*, which works out as

$$\mathrm{SSD}_{\mathrm{res}} = \sum_i \sum_j (x_{ij} - \bar{x}_{i.} - \bar{x}_{.j} + \bar{x}_{..})^2$$

This corresponds to a statistical model in which it is assumed that the observations are composed of a general level, a row effect, and a column effect plus a noise term:

$$X_{ij} = \mu + \alpha_i + \beta_j + \epsilon_{ij} \qquad \epsilon_{ij} \sim N(0, \sigma^2)$$

The parameters of this model are not uniquely defined unless we impose some restriction on the parameters. If we impose $\sum \alpha_i = 0$ and $\sum \beta_j = 0$, then the estimates of α_i, β_j, and μ turn out to be $\bar{x}_{i\cdot} - \bar{x}_{\cdot\cdot}$, $\bar{x}_{\cdot j} - \bar{x}_{\cdot\cdot}$, and $\bar{x}_{\cdot\cdot}$.

Dividing the sums of squares by their respective degrees of freedom $m - 1$ for SSD_R, $n - 1$ for SSD_C, and $(m - 1)(n - 1)$ for SSD_{res}, we get a set of mean squares. F tests for no row and column effect can be carried out by dividing the respective mean squares with the residual mean square.

It is important to notice that this works out so nicely only because of the balanced design. If you have a table with "holes" in it, the analysis is considerably more complicated. The simple formulas for the sum of squares are no longer valid and, in particular, the order independence is lost, so that there is no longer a single SSD_C, but one with and without adjusting for row effects.

To perform a two-way ANOVA, it is necessary to have data in one vector, with the two classifying factors parallel to it. We consider an example concerning heart rate after administration of enalaprilate (Altman, 1991, p. 327). Data are found in this form in the heart.rate data set:

```
> data(heart.rate)
> attach(heart.rate)
> heart.rate
      hr subj time
1    96     1     0
2   110     2     0
3    89     3     0
4    95     4     0
5   128     5     0
6   100     6     0
7    72     7     0
8    79     8     0
9   100     9     0
10   92     1    30
11  106     2    30
12   86     3    30
13   78     4    30
14  124     5    30
15   98     6    30
16   68     7    30
17   75     8    30
18  106     9    30
```

```
19   86    1    60
20  108    2    60
21   85    3    60
22   78    4    60
23  118    5    60
24  100    6    60
25   67    7    60
26   74    8    60
27  104    9    60
28   92    1   120
29  114    2   120
30   83    3   120
31   83    4   120
32  118    5   120
33   94    6   120
34   71    7   120
35   74    8   120
36  102    9   120
```

If you look inside the heart.rate.R file in the data directory of the
ISwR package, you will see that the actual definition of the data frame is

```
heart.rate <- data.frame(hr = c(96,110,89,95,128,100,72,79,100,
                                92,106,86,78,124, 98,68,75,106,
                                86,108,85,78,118,100,67,74,104,
                                92,114,83,83,118,94,71,74,102),
                         subj=gl(9,1,36),
                         time=gl(4,9,36,labels=c(0,30,60,120))))
```

The gl function (generate *levels*) is specially designed for generating pat-
terned factors for balanced experimental designs. It has three arguments:
the number of levels, the block length (how many times each level should
repeat), and the total length of the result. The two patterns in the data
frame are thus

```
> gl(9,1,36)
 [1] 1 2 3 4 5 6 7 8 9 1 2 3 4 5 6 7 8 9 1 2 3 4 5 6 7 8 9 1 2 3
[31] 4 5 6 7 8 9
Levels:  1 2 3 4 5 6 7 8 9
> gl(4,9,36,labels=c(0,30,60,120))
 [1] 0    0    0    0    0    0    0    0    0    30   30   30   30   30   30
[16] 30   30   30   60   60   60   60   60   60   60   60   60   120 120 120
[31] 120 120 120 120 120 120
Levels:  0 30 60 120
```

Once the variables have been defined, the two-way analysis of variance is
specified simply by

```
> anova(lm(hr~subj+time))
Analysis of Variance Table
```

```
Response: hr
            Df Sum Sq Mean Sq F value    Pr(>F)
subj         8 8966.6  1120.8 90.6391 4.863e-16 ***
time         3  151.0    50.3  4.0696   0.01802 *
Residuals   24  296.8    12.4
---
Signif. codes:  0 '***' 0.001 '**' 0.01 '*' 0.05 '.' 0.1 ' ' 1
```

Interchanging `subj` and `time` in the model formula (`hr~time+subj`) yields exactly the same analysis except for the order of the rows of the ANOVA table. This is because we are dealing with a balanced design (a complete two-way table with no missing values). In unbalanced cases the factor order will matter.

6.3.1 Graphics for repeated measurements

At least for your own use, it is useful to plot a "spaghettigram" of the data, that is, a plot where data from the same subject are connected with lines. To this end, you can use the function `interaction.plot`, which graphs the values against one factor, while connecting data for the other factor with line segments to form traces.

```
> interaction.plot(time, subj, hr)
```

In fact there is a fourth argument, which specifies what should be done in case, there is more than one observation per cell. By default, the mean is taken, which is the reason why the y-axis in Figure 6.2 reads "mean of hr".

If you prefer to have the values plotted according to the time of measurement (which are not equidistant in this example), you could instead write (resulting plot not shown)

```
> interaction.plot(ordered(time),subj,hr)
```

6.4   The Friedman test

A nonparametric counterpart of two-way analysis of variance exists for the case with one observation per cell. Friedman's test is based on ranking observations *within each row* assuming that if there is no column effect then all orderings should be equally likely. A test statistic based on the column sum of squares can be calculated and normalized to give a χ^2-distributed test statistic.

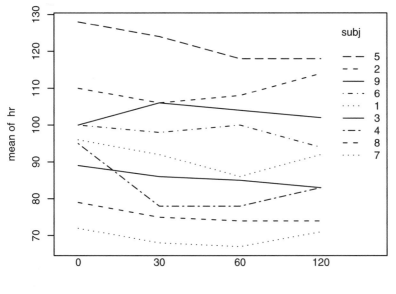

Figure 6.2. Interaction plot of heart-rate data.

In the case of two columns the Friedman test is equivalent to the *sign test*, in which one uses the binomial distribution to test for equal probability of positive and negative differences within pairs. This is a rather less sensitive test than the Wilcoxon signed-rank test discussed in Section 4.2.

Practical application of the test is as follows:

```
> friedman.test(hr~time|subj,data=heart.rate)

        Friedman rank sum test

data:  hr and time and subj
Friedman chi-squared = 8.5059, df = 3, p-value = 0.03664
```

Notice that the blocking factor is specified in a model formula using the vertical bar, which may be read as "time within subj". It is seen that the test is not quite as strongly significant as the parametric counterpart. This is unsurprising since the latter test is more powerful when its assumptions are met.

6.5   The ANOVA table in regression analysis

We have seen the use of analysis of variance tables in grouped and cross-classified experimental designs. However, their use is not restricted to these designs but applies to the whole class of *linear models* (more on this in Chapter 10).

The variation between and within groups for a one-way analysis of variance generalizes to *model variation* and *residual variation*

$$\text{SSD}_{\text{model}} = \sum_i (\hat{y}_i - \bar{y}.)^2$$

$$\text{SSD}_{\text{res}} = \sum_i (y_i - \hat{y}_i)^2$$

which partition the total variation $\sum_i (y_i - \bar{y}.)^2$. This applies only when the model contains an intercept; see Section 10.2. The role of the group means in the one-way classification is taken over by the fitted values $\hat{y}_i$ in the more general linear model.

An F test for significance of the model is available in direct analogy with Section 6.1. In simple linear regression this test is equivalent to testing that the regression coefficient is zero.

The analysis of variance table corresponding to a regression analysis can be extracted with the function anova, just as for one- and two-way analyses of variance. For the thuesen example, it will look like this:

```
> data(thuesen)
> attach(thuesen)
> lm.velo <- lm(short.velocity~blood.glucose)
> anova(lm.velo)
Analysis of Variance Table

Response: short.velocity
              Df  Sum Sq Mean Sq F value Pr(>F)
blood.glucose  1 0.20727 0.20727   4.414 0.0479 *
Residuals     21 0.98610 0.04696
---
Signif. codes:  0 `***' 0.001 `**' 0.01 `*' 0.05 `.' 0.1 ` ' 1
```

Notice that the F test gives the same p-value as the t test for a zero slope from Section 5.1. It is the same F test that gets printed at the end of the summary output:

```
...
Residual standard error: 0.2167 on 21 degrees of freedom
Multiple R-Squared: 0.1737,     Adjusted R-squared: 0.1343
F-statistic: 4.414 on 1 and 21 DF, p-value: 0.0479
```

The remaining elements of the three output lines above may also be derived from the ANOVA table. "Residual standard error" is the square root of "Residual mean squares", namely $0.2167 = \sqrt{0.04696}$. R^2 is the proportion of the total sum of squares explained by the regression line, that is, $0.1737 = 0.2073/(0.2073 + 0.9861)$; and, finally, the adjusted R^2 is the relative improvement of the residual variance, $0.1343 = (v - 0.04696)/v$, where $v = (0.2073 + 0.9861)/22 = 0.05425$ is the variance of short.velocity if the glucose values are not taken into account.

6.6 Exercises

6.1 The zelazo data are in the form of a list of vectors, one for each of the four groups. Convert the data to a form suitable for the use of lm, and calculate the relevant test. Consider t tests comparing selected subgroups or obtained by combining groups.

6.2 In the lung data do the three measurement methods give systematically different results? If so, which ones appear to be different?

6.3 Repeat the previous exercises using the zelazo and lung data with the relevant nonparametric tests.

6.4 The igf1 variable in the juul data set is arguably skewed and has different variance across Tanner groups. Try to compensate for this using logarithmic and square-root transformations, and use the Welch test. However, the analysis is still problematic — why?

7

Tabular data

This chapter describes a series of functions designed to analyze tabular data. Specifically, we look at the functions `prop.test`, `binom.test`, `chisq.test`, and `fisher.test`.

7.1 Single proportions

Tests of single proportions are generally based on the binomial distribution (see Section 2.3) with size parameter N and probability parameter p. For large sample sizes, this can be well approximated by a normal distribution with mean Np and variance $Np(1 - p)$. As a rule of thumb, the approximation is satisfactory when the expected number of "successes" and "failures" are both larger than 5.

Denoting the observed number of "successes" by x, the test for the hypothesis that $p = p_0$ can be based on

$$u = \frac{x - Np_0}{\sqrt{Np_0(1 - p_0)}}$$

which has an approximate normal distribution with mean zero and standard deviation 1, or on u^2, which has an approximate χ^2 distribution with 1 degree of freedom.

The normal approximation can be somewhat improved by the *Yates correction*, which shrinks the observed value by half a unit toward the expected value when calculating u.

We consider an example (Altman, 1991, p. 230) where 39 of 215 randomly chosen patients are observed to have asthma, and one wants to test the hypothesis that the probability of a "random patient" having asthma is 0.15. This can be done using prop.test:

```
> prop.test(39,215,.15)

        1-sample proportions test with continuity correction

data:   39 out of 215, null probability 0.15
X-squared = 1.425, df = 1, p-value = 0.2326
alternative hypothesis: true p is not equal to 0.15
95 percent confidence interval:
 0.1335937 0.2408799
sample estimates:
        p
0.1813953
```

The three arguments to prop.test are the number of positive outcomes, the total number, and the (theoretical) probability parameter that you want to test for. The latter is 0.5 by default, which makes sense for symmetrical problems, but this is not the case here. The amount 15% is a bit synthetic since it is rarely the case that one has a specific a priori value to test for. It is usually more interesting to compute a confidence interval for the probability parameter, such as is given in the last part of the output. Notice that we have a slightly unfortunate double usage of the symbol p as the probability parameter of the binomial distribution and as the test probability or p-value.

You can also use binom.test to obtain a test in the binomial distribution. In that way you get an exact test probability, so it is generally preferable to using prop.test, but prop.test can do more than testing single proportions. The procedure to obtain the p-value is to calculate the point probabilities for all the possible values of x and sum those that are less than or equal to the point probability of the observed x.

```
> binom.test(39,215,.15)

        Exact binomial test

data:   39 and 215
number of successes = 39, number of trials = 215, p-value = 0.2135
alternative hypothesis: true probability  ...   not equal to 0.15
95 percent confidence interval:
 0.1322842 0.2395223
sample estimates:
```

```
probability of success
          0.1813953
```

The "exact" confidence intervals at the 0.05 level are actually constructed from the two one-sided tests at the 0.025 level. Finding an exact confidence interval using two-sided tests is not a well-defined problem (see Exercise 7.5).

7.2 Two independent proportions

The function prop.test can also be used to compare two or more proportions. For that purpose, the arguments should be given as two vectors, where the first contains the numbers of positive outcomes and the second the total numbers for each group.

The theory is similar to that for a single proportion: Consider the difference in the two proportions $d = x_1/N_1 - x_2/N_2$, which will be approximately normally distributed with mean zero and variance $V_p(d) = (1/N_1 + 1/N_2) * p(1 - p)$ if the counts are binomially distributed with the same p parameter. So to test the hypothesis that $p_1 = p_2$, plug the common estimate $\hat{p} = (x_1 + x_2)/(n_1 + n_2)$ into the variance formula and look at $u = d/\sqrt{V_{\hat{p}}(d)}$, which approximately follows a standard normal distribution, or look at u^2, which is approximately $\chi^2(1)$-distributed. A Yates-type correction is possible, but we skip the details.

For illustration, we use an example originally due to Lewitt and Machin (Altman, 1991, p. 232):

```
> lewitt.machin.success <- c(9,4)
> lewitt.machin.total <- c(12,13)
> prop.test(lewitt.machin.success,lewitt.machin.total)

        2-sample test for equality of proportions with
        continuity correction

data:  lewitt.machin.success out of lewitt.machin.total
X-squared = 3.2793, df = 1, p-value = 0.07016
alternative hypothesis: two.sided
95 percent confidence interval:
 0.01151032 0.87310506
sample estimates:
   prop 1    prop 2
0.7500000 0.3076923
```

The confidence interval given is for the *difference* in proportions. The theory behind its calculation is similar to that of the test, but there are some technical complications and a different approximation is used.

You can also perform the test without the Yates continuity correction. This is done by adding the argument `correct=F`. The continuity correction makes the confidence interval somewhat wider than it would otherwise be, but notice that it nevertheless does not contain zero. Thus, the confidence interval is contradicting the test, which says that there is *no* significant difference between the two groups with a two-sided test. The explanation lies in the different approximations, which becomes important for tables as sparse as the present one.

If you want at least to be sure that the *p*-value is correct, you can use Fisher's exact test. We illustrate this using the same data as in the preceding section. The test works by making the calculations in the conditional distribution of the 2 × 2 table given both the row and column marginals. This can be difficult to envision, but think of it like this: Take 13 white balls and 12 black balls (success and failure, respectively), and sample the balls without replacement into two groups of size 12 and 13. The number of white balls in the first group obviously defines the whole table, and the point is that its distribution can be found as a purely combinatorial problem. The distribution is known as the *hypergeometric distribution*.

The relevant function is `fisher.test`, which requires data to be given in matrix form. This is obtained as follows:

```
> matrix(c(9,4,3,9),2)
      [,1] [,2]
[1,]    9    3
[2,]    4    9

> lewitt.machin <- matrix(c(9,4,3,9),2)
> fisher.test(lewitt.machin)

        Fisher's Exact Test for Count Data

data:   lewitt.machin
p-value = 0.04718
alternative hypothesis: true odds ratio is not equal to 1
95 percent confidence interval:
  0.9006803 57.2549701
sample estimates:
odds ratio
  6.180528
```

Notice that the second column of the table needs to be the number of negative outcomes, not the total number of observations.

Notice also that the confidence interval is for the *odds ratio*, that is, for the estimate of $(p_1/(1 - p_1))/(p_2/(1 - p_2))$. One can show that if the *p*s are not identical, then the conditional distribution of the table depends only on the odds ratio, so it is the natural measure of association in connection with the Fisher test. The exact distribution of the test can be worked out also when the odds ratio differs from 1, but there is the same complication as with binom.test that a two-sided 95% confidence interval must be pasted together from two one-sided 97.5% intervals. This leads to the opposite inconsistency as with prop.test: The test is (barely) significant, but the confidence interval for the odds ratio includes 1.

The standard χ^2 test (see also Section 7.4) in chisq.test works with data in matrix form, like fisher.test does. For a 2 × 2 table, the test is exactly equivalent to prop.test.

```
> chisq.test(lewitt.machin)

        Pearson's Chi-squared test with Yates' continuity
        correction

data:   lewitt.machin
X-squared = 3.2793, df = 1, p-value = 0.07016
```

7.3   *k* proportions, test for trend

Sometimes you want to compare more than two proportions. In that case the categories are often ordered so that you would expect to find a decreasing or increasing trend in the proportions with the group number.

This example used in this section concerns data from a group of women giving birth where it was recorded whether the child was delivered by Caesarean section and what shoe size the mother used (Altman, 1991, p. 229). The data are given as a table and can be entered as follows:

```
> caesar.shoe <- matrix(c(5,7,6,7,8,10,
+                         17,28,36,41,46,140), nrow=2, byrow=T)
> colnames(caesar.shoe) <- c("<4","4","4.5","5","5.5","6+")
> rownames(caesar.shoe) <- c("Yes","No")
```

The data can also be loaded using data(caesarean). Either way, we obtain this table:

```
> caesar.shoe
     <4  4 4.5  5 5.5  6+
Yes   5  7   6  7   8  10
No   17 28  36 41  46 140
```

To compare $k > 2$ proportions, another test based on the normal approximation is available. It consists of the calculation of a weighted sum of squared deviations between the observed proportions in each group and the overall proportion for all groups. The test statistic has an approximate χ^2 distribution with $k - 1$ degrees of freedom.

To use prop.test on a table like caesar.shoe, we need to convert it to a vector of "successes" (which in this case is close to being the opposite) and a vector of "trials". The two vectors can be computed like this:

```
> caesar.shoe.yes <- caesar.shoe["Yes",]
> caesar.shoe.total <- margin.table(caesar.shoe,2)
> caesar.shoe.yes
 <4   4 4.5   5 5.5  6+
  5   7   6   7   8  10
> caesar.shoe.total
 <4   4 4.5   5 5.5  6+
 22  35  42  48  54 150
```

Thereafter it is easy to perform the test:

```
> prop.test(caesar.shoe.yes,caesar.shoe.total)
        6-sample test for equality of proportions without
        continuity correction

data:  caesar.shoe.yes out of caesar.shoe.total
X-squared = 9.2874, df = 5, p-value = 0.09814
alternative hypothesis: two.sided
sample estimates:
     prop 1     prop 2     prop 3     prop 4     prop 5
0.22727273 0.20000000 0.14285714 0.14583333 0.14814815
     prop 6
0.06666667

Warning message:
Chi-squared approximation may be incorrect in: prop.test(...
```

It is seen that the test comes out nonsignificant, but the subdivision is really unreasonably fine in view of the small number of Caesarean sections. Notice, by the way, the warning about the χ^2 approximation being dubious, which is prompted by some cells having an expected count less than 5.

You can test for a trend in the proportions using prop.trend.test. It takes three arguments: x, n, and score. The first two of these are exactly like in prop.test, whereas the last one is the score given to the groups, by default simply $1, 2, \ldots, k$. The basis of the the test is essentially a weighted linear regression of the proportions on the group scores, where we test for a zero slope, which becomes a χ^2 test on 1 degree of freedom.

```
> prop.trend.test(caesar.shoe.yes,caesar.shoe.total)

        Chi-squared Test for Trend in Proportions

data:  caesar.shoe.yes out of caesar.shoe.total ,
 using scores: 1 2 3 4 5 6
X-squared = 8.0237, df = 1, p-value = 0.004617
```

So if we assume that the effect of shoe size is linear in the group score, *then* we can see a significant difference. This kind of assumption should not be thought of as something that must hold for the test to be valid. Rather, it indicates the rough type of alternative to which the test should be sensitive.

The effect of using a trend test can be viewed as an approximate subdivision of the test for equal proportions ($\chi^2 = 9.29$) into a contribution from the linear effect ($\chi^2 = 8.02$) on 1 degree of freedom and a contribution from deviations from the linear trend ($\chi^2 = 1.27$) on 4 degrees of freedom. So you could say that the test for equal proportions is being diluted or wastes degrees on freedom on testing for deviations in a direction we are not really interested in.

7.4 $r \times c$ tables

For the analysis of tables with more than two classes on both sides, you can use chisq.test or fisher.test although you should note that the latter can be very computationally demanding if the cell counts are large and there are more than two rows or columns. We have already seen chisq.test in a simple example, but with larger tables, some additional features are of interest.

An $r \times c$ table looks like this:

$$
\begin{array}{cccc|c}
n_{11} & n_{12} & \cdots & n_{1c} & n_{1.} \\
n_{21} & n_{22} & \cdots & n_{2c} & n_{2.} \\
\vdots & \vdots & & \vdots & \vdots \\
n_{r1} & n_{r2} & \cdots & n_{rc} & n_{r.} \\
\hline
n_{.1} & n_{.2} & \cdots & n_{.c} & n_{..}
\end{array}
$$

Such a table can arise from several different sampling plans, and the notion of "no relation between rows and columns" is correspondingly different. The total in each row might be fixed in advance and you would be interested in testing whether the distribution over columns is the same for each row, or vice versa if the column totals were fixed. It might also be the case that only the total number is chosen and the individuals are grouped randomly according to the row and column criteria. In the latter case you would be interested in testing the hypothesis of *statistical independence*, that the probability of an individual falling into the ijth cell is the product $p_{i.}p_{.j}$ of the marginal probabilities. However, the analysis of the table turns out to be the same in all cases.

If there is no relation between rows and columns, then you would expect to have the following cell values:

$$E_{ij} = \frac{n_{i.} \times n_{.j}}{n_{..}}$$

This can be interpreted as distributing each row total according to the proportions in each column (or vice versa) or as distributing the grand total according to the products of the row and column proportions.

The test statistic

$$X^2 = \sum \frac{(O - E)^2}{E}$$

has an approximate χ^2 distribution with $(r - 1) \times (c - 1)$ degrees of freedom. Here the sum is over the entire table and the ij indices have been omitted. O denotes the observed values and E the expected values as described above.

We consider the table with caffeine consumption and marital status from Section 3.5 and compute the χ^2 test:

```
> caff.marital <- matrix(c(652,1537,598,242,36,46,38,21,218
+ ,327,106,67),
+ nrow=3,byrow=T)
> colnames(caff.marital) <- c("0","1-150","151-300",">300")
> rownames(caff.marital) <- c("Married","Prev.married","Single")
> caff.marital
                0 1-150 151-300 >300
Married       652  1537     598  242
Prev.married   36    46      38   21
Single        218   327     106   67
> chisq.test(caff.marital)

        Pearson's Chi-squared test

data:  caff.marital
X-squared = 51.6556, df = 6, p-value = 2.187e-09
```

The test is highly significant, so we can safely conclude that the data contradict the hypothesis of independence. However, you would generally also like to know the nature of the deviations. To that end, you can look at some extra components of the return value of chisq.test.

Notice that chisq.test (just like lm) actually returns more information than what is commonly printed:

```
> chisq.test(caff.marital)$expected
                    0      1-150    151-300       >300
Married       705.83179 1488.01183 578.06533 257.09105
Prev.married   32.85648   69.26698  26.90895  11.96759
Single        167.31173  352.72119 137.02572  60.94136
> chisq.test(caff.marital)$observed
               0 1-150 151-300 >300
Married      652  1537     598  242
Prev.married  36    46      38   21
Single       218   327     106   67
```

These two tables may then be scrutinized to see wherein the differences lie. It is often useful to look at a table of the contributions from each cell to the total χ^2. Such a table cannot be directly extracted, but it is easy to calculate:

```
> E <- chisq.test(caff.marital)$expected
> O <- chisq.test(caff.marital)$observed
> (O-E)^2/E
                     0      1-150    151-300      >300
Married       4.1055981 1.612783 0.6874502 0.8858331
Prev.married  0.3007537 7.815444 4.5713926 6.8171090
Single       15.3563704 1.875645 7.0249243 0.6023355
```

There are some large contributions, particularly from too many "abstaining" singles, and the distribution among previously married is shifted in the direction of a larger intake — insofar as they consume caffeine at all. Still, it is not easy to find a simple description of the deviation from independence in these data.

You can also use chisq.test directly on raw (untabulated) data, here using the juul data set from Section 3.5:

```
> data(juul)
> attach(juul)
> chisq.test(tanner,sex)

        Pearson's Chi-squared test

data:  tanner and sex
X-squared = 28.8672, df = 4, p-value = 8.318e-06
```

It may not really be relevant to test for independence between these particular variables. The definition of Tanner stages is gender-dependent by nature.

7.5   Exercises

7.1 Reconsider the situation of Exercise 2.3 where 10 consecutive patients had operations without complications and the expected rate was 20%. Calculate the relevant one-sided test in the binomial distribution. How large a sample (still with zero complications) would be necessary to obtain statistical significance?

7.2 In 747 cases of "Rocky Mountain spotted fever" from the western United States, 210 patients died. Out of 661 cases from the eastern United States, 122 died. Is the difference statistically significant? (See also Exercise 11.4.)

7.3 Two drugs for the treatment of peptic ulcer were compared (Campbell and Machin, 1993, p. 72). The results were as follows:

	Healed	Not healed	Total
Pirenzepine	23	7	30
Trithiozine	18	13	31
Total	41	20	61

Compute the χ^2 test and Fisher's exact test and discuss the difference. Find an approximate 95% confidence interval for the difference in healing probability.

7.4 (From "Mathematics 5" exam, U. Copenhagen, Summer 1969) From September 20, 1968, to February 1, 1969, an instructor consumed 254 eggs. Every day, he recorded how many eggs broke during the boiling so that the white ran out, and how many cracked so that the white did not run out. Additionally, he recorded whether it was size A eggs or size B. From February 4, 1969, until April 10, 1969, he consumed 130 eggs, but this time he used a "piercer" for creating a small hole in the egg to prevent breaking and cracking. The results were as follows:

Period	Size	Total	Broken	Cracked
Sept. 20–Feb. 1	A	54	4	8
Sept. 20–Feb. 1	B	200	15	28
Feb. 4–Apr. 10	A	60	4	9
Feb. 4–Apr. 10	B	70	1	7

Investigate whether or not the piercer seems to have had an effect.

7.5 Make a plot of the two-sided p value for testing that the probability parameter is x when the observations are 3 successes in 15 trials, for x varying from 0 to 1 in steps of 0.001. Explain what makes the definition of a two-sided confidence interval difficult.

8

Power and the computation of sample size

A statistical test will not be able to detect a true difference if the sample size is too small compared to the magnitude of the difference. When designing experiments, the experimenter should try to ensure that a sufficient amount of data is collected to be reasonably sure that a difference of a specified size will be detected. R has methods for doing these calculations in the simple cases of comparing means using one- or two-sample t tests and comparing two proportions.

8.1 The principles of power calculations

This section outlines the theory of power calculations and sample-size choice. If you are practically inclined and just need to find the necessary sample size in a particular situation, you can safely skim this section and move quickly to subsequent sections that contain the actual R calls.

The basic idea of a hypothesis test should be clear by now. A test statistic is defined and its value is used to decide whether or not you can accept the (null) hypothesis. Acceptance and rejection regions are set up so that the probability of getting a test statistic that falls into the rejection region is a specified significance level (α) if the null hypothesis is true. In the present context it is useful to stick to this formulation (as opposed to the use of p values), as rigid as it might be.

Since data are sampled at random, there is always a risk of reaching a wrong conclusion, and things can go wrong in two ways:

- The hypothesis is correct, but the test rejects it (type I error).
- The hypothesis is wrong, but the test accepts it (type II error).

The risk of a type I error is the significance level. The risk of a type II error will depend on the size and nature of the deviation you are trying to detect. If there is very little difference, then you do not have much of a chance of detecting it. For this reason, some statisticians disapprove of terms like "acceptance region" because you can never prove that there is no difference — you can only not prove that there is one.

The probability of rejecting a false hypothesis is called the *power* of the test, and methods exist for calculating or approximating the power in the most important practical situations. It is inconvenient to talk further about these matters in the abstract, so let's move on to some concrete examples.

8.1.1 The power of one-sample and paired t tests

Consider the case of the comparison of a sample mean to a given value. For example, in a matched trial we wish to test whether the difference between treatment A and treatment B is zero using a paired t test (described in Chapter 4).

We call the true difference δ. Even if the null hypothesis is not true, we can still work out the distribution of the test statistic, provided the other model assumptions hold. It is called the *noncentral t distribution* and depends on a noncentrality parameter as well as the usual degrees of freedom. For the paired t test, the noncentrality parameter ν is a function of δ, the standard deviation of differences σ, and the sample size n and equals

$$\nu = \frac{\delta}{\sigma/\sqrt{n}}$$

that is, it is simply the true difference divided by the standard error of the mean.

The cumulative noncentral t distribution is available in R simply by adding an ncp argument to the pt function. As of version 1.5.0, the density is not available. Figure 8.1 shows a plot of pt with ncp=3 and df=25. In the same figure, a vertical line indicates the upper end of the acceptance region for a two-sided test at the 0.05 significance level. The plot was created as follows:

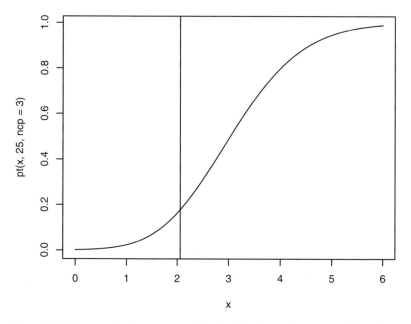

Figure 8.1. The cumulative noncentral t distribution with $\nu = 3$ and 25 degrees of freedom. The vertical line marks the upper significance limit for a two-sided test at the 0.05 level.

```
> curve(pt(x,25,ncp=3), from=0, to=6)
> abline(v=qt(.975,25))
```

The plot shows the main part of the distribution falling in the rejection region. The probability of getting a value in the acceptance region can be seen from the graph as the intersection between the curve and the vertical line. (Almost! See Exercise 8.4.) This value is easily calculated as

```
> pt(qt(.975,25),25,ncp=3)
[1] 0.1779891
```

or roughly 0.18. The power of the test is the opposite, the probability of getting a significant result. In this case it is 0.82, and it is of course desirable to have the power as close to 1 as possible.

Notice that the power (traditionally denoted β) depends on four quantities: δ, σ, n, and α. If we fix any three of these, we can adjust the fourth to achieve a given power. This can be used to determine the necessary sample size for an experiment: You need to specify a desired power ($\beta = 0.80$ and $\beta = 0.90$ are common choices), the significance level (usually given by convention as $\alpha = 0.05$), a guess of the standard deviation, and δ, which is known as the "minimal relevant difference" (MIREDIF) or "smallest meaningful difference" (SMD). This gives an equation that you can solve for n. The result will generally be a fractional number, which should of course be rounded up.

You can also work on the opposite problem and answer the following question: Given a feasible sample size, how large a difference should you reasonably be able to detect?

Sometimes a shortcut is made by expressing δ relative to the standard deviation, in which case you would simply set σ to 1.

8.1.2 Power of two-sample t test

Procedures for two-sample t tests are essentially the same as for the one-sample case, except for the calculation of the noncentrality parameter, which is calculated as

$$\nu = \frac{\delta}{\sigma\sqrt{1/n_1 + 1/n_2}}$$

It is generally assumed that the variance is the same in the two groups, that is, using the Welch procedure is not considered. In sample-size calculations, one usually assumes that the group sizes are the same, since that gives the optimal power for a given total number of observations.

8.1.3 Approximate methods

For hand calculations, the power calculations can be considerably simplified by assuming that the standard deviation is known, so that the t test is replaced by a test in the standard normal distribution. The practical advantage is that the approximate formula for the power is easily inverted to give an explicit formula for n. For the one- and two-sample cases, this works out as

$$n = \left(\frac{\Phi_{\alpha/2} + \Phi_{\beta}}{\delta/\sigma}\right)^2 \qquad \text{one-sample}$$

$$n = 2 \times \left(\frac{\Phi_{\alpha/2} + \Phi_{\beta}}{\delta/\sigma}\right)^2 \qquad \text{two-sample, each group}$$

with the Φ_x denoting quantiles on the normal distribution. This is for two-sided tests. For one-sided tests, use α instead of $\alpha/2$.

These formulas are often found in textbooks, and some computer programs implement them rather than the more accurate method described earlier. They do have the advantage of more clearly displaying theoretical properties like the proportionality of δ and $1/\sqrt{n}$ for a given power. However, they become numerically unsatisfactory when the degrees of freedom falls below 20 or so.

8.1.4 Power of comparisons of proportions

Suppose you wish to compare the morbidity between two populations and have to decide the number of persons to sample from each population. That is, you plan to perform a comparison of two binomial distributions as described in Section 7.2 using `prop.test` or `chisq.test`.

For binomial comparisons exact power calculations become unwieldy, so we rely on normal approximations to the binomial distribution. The power will depend on the probability in both groups, not just their difference. As for the t test, the group sizes are assumed to be equal. The theoretical derivation of the power proceeds along the same lines as before, by calculating the distribution of $\hat{p}_1 - \hat{p}_2$ when $p_1 \neq p_2$ and the probability that it falls outside the range of values compatible with the hypothesis $p_1 = p_2$. Assuming equal numbers in the two groups, this leads to the sample-size formula

$$n = \left(\frac{\Phi_{\alpha/2}\sqrt{2p(1-p)} + \Phi_\beta \sqrt{p_1(1-p_1) + p_2(1-p_2)}}{|p_2 - p_1|} \right)^2$$

in which $p = (p_1 + p_2)/2$.

Since the method is only approximate, the results are not reliable unless the expected number in each of the four cells in the 2×2 table is greater than 5.

8.2 Two-sample problems

The following example is from Altman (1991, p. 457) and concerns the influence of milk on growth. Two groups are to be given different diets, and their growth will be measured. We wish to compute the sample size that with a power of 90%, using a two-sided test at the 1% level, can find a difference of 0.5 cm in a distribution with a standard deviation of 2 cm. This is done as follows:

```
> power.t.test(delta=0.5, sd=2, sig.level = 0.01, power=0.9)

        Two-sample t test power calculation

              n = 477.8021
          delta = 0.5
             sd = 2
      sig.level = 0.01
          power = 0.9
    alternative = two.sided

NOTE: n is number in *each* group
```

delta stands for the "true difference", and sd is the standard deviation. As is seen, the calculation may return a fractional number of experimental unit. This would, of course, in practice be rounded up to 478. In the original reference, a method employing nomograms (a graphical technique) is used and the value obtained is 450. The difference is probably due to difficulty in reading the value off the nomogram scale. To know which power you would actually obtain with 450 in each group, you would enter

```
> power.t.test(n=450, delta=0.5, sd=2, sig.level = 0.01)

        Two-sample t test power calculation

              n = 450
          delta = 0.5
             sd = 2
      sig.level = 0.01
          power = 0.8784433
    alternative = two.sided

NOTE: n is number in *each* group
```

The system is that exactly four out of five arguments (power, sig.level, delta, sd, and n) are given, whereafter the function computes the missing one (defaults exist to set sd=1 and sig.level=0.05 — if you wish to have those calculated, then you must explicitly pass them as NULL). In addition, there are two optional arguments: alternative, which can be used to specify one-sided tests; and type, which can be used to specify that you want to handle a one-sample problem. The former is used like this (recall that the + is R's secondary prompt indicating that it is waiting for the remainder of an unfinished command):

```
> power.t.test(delta=0.5, sd=2, sig.level = 0.01, power=0.9,
+ alt="one.sided")
```

```
Two-sample t test power calculation

            n = 417.898
        delta = 0.5
           sd = 2
    sig.level = 0.01
        power = 0.9
  alternative = one.sided

NOTE: n is number in *each* group
```

8.3   One-sample problems and paired tests

One-sample problems are handled by adding `type="one.sample"` in the call to `power.t.test`. Similarly, paired tests are specified with `type="paired"`; although these reduce to one-sample tests by forming differences, the printout will be slightly different.

One pitfall when planning a study with paired data is that the literature sometimes gives the intra-individual variation as "standard deviation of repeated measurements on the same person" or similar. These may be calculated by measuring a number of persons several times and computing a common standard deviation within persons. This needs to be multiplied by $\sqrt{2}$ to get the standard deviation of differences, which `power.t.test` requires for paired data. If, for instance, it is known that the standard deviation within persons is about 10, and you want to use a paired test at the 5% level to detect a difference of 10 with a power of 85%, then you should enter

```
> power.t.test(delta=10, sd=10*sqrt(2), power=0.85, type="paired")

        Paired t test power calculation

            n = 19.96892
        delta = 10
           sd = 14.14214
    sig.level = 0.05
        power = 0.85
  alternative = two.sided

NOTE: n is number of *pairs*, sd is std.dev. of
      *differences* within pairs
```

Notice that `sig.level=0.05` was taken as the default.

8.4   Comparison of proportions

To calculate sample sizes and related quantities for comparisons of proportions, you should use power.prop.test. This is based on approximations with the normal distribution, so do not trust the results if any of the expected cell counts drop below 5.

The use of power.prop.test is analogous to power.t.test, although delta and sd are replaced by the hypothesized probabilities in the two groups, p1 and p2. Currently, it is not possible to specify that one wants to consider a one-sample problem.

An example is given in Altman (1991, p. 459) in which two groups are administered or not administered nicotine chewing gum and the binary outcome is smoking cessation. The stipulated values are $p_1 = 0.15$ and $p_2 = 0.30$. We want a power of 85%, and the significance level is the traditional 5%. Inserting these values yields

```
> power.prop.test(power=.85,p1=.15,p2=.30)

        Two-sample comparison of proportions power calculation

              n = 137.6040
             p1 = 0.15
             p2 = 0.3
      sig.level = 0.05
          power = 0.85
    alternative = two.sided

 NOTE: n is number in *each* group
```

8.5   Exercises

8.1   The ashina trial was designed to have 80% power if the true treatment difference was 15% and the standard deviation of differences within a person was 20%. Comment on the sample size chosen. (The power calculation was originally done using the approximative formula. The imbalance between the group sizes is due to the use of an open randomization procedure.)

8.2   In a trial comparing a binary outcome between two groups, find the required number of patients to find an increase in the success rate from 60% to 75% with a power of 90%. What happens if we reduce the power requirement to 80%?

8.3   (Theoretical.) Even though dt does not allow the ncp argument, it is quite easy to get the approximate density of the noncentral t distribution by numerical differentiation of the pt function. Plot the density for ncp=3 and df=25 and compare it to the distribution of $t + 3$, where t has a central t distribution with df=25.

8.4 In two-sided tests there is also a risk of falling into the rejection region on the opposite side of the true value. The power calculations in R do not take this into account. Discuss the consequences.

8.5 It is occasionally suggested to choose n to "make the true difference significant". What power would result from choosing n by such a procedure?

9

Multiple regression

This chapter discusses the case of regression analysis with multiple predictors. There is not really much new here, since model specification and output do not differ a lot from what has been described for regression analysis and analysis of variance. The news is mainly the model search aspect, namely, among a set of potential descriptive variables to look for a subset that describes the response sufficiently well.

The basic model for multiple regression analysis is

$$y = \beta_0 + \beta_1 x_1 + \cdots + \beta_k x_k + \epsilon$$

where $x_1, \ldots x_k$ are explanatory variables (also called predictors) and the parameters $\beta_1, \ldots, \beta_k$ can be estimated using the method of least squares (see Section 5.1). A closed-form expression for the estimates can be derived using matrix calculus, but we do not go into details with that here.

9.1 Plotting multivariate data

As an example in this chapter, we use a study concerning lung function in patients with cystic fibrosis in Altman (1991, p. 338).

Data are in the ISwR package and can be loaded into the workspace with

```
> data(cystfibr)
```

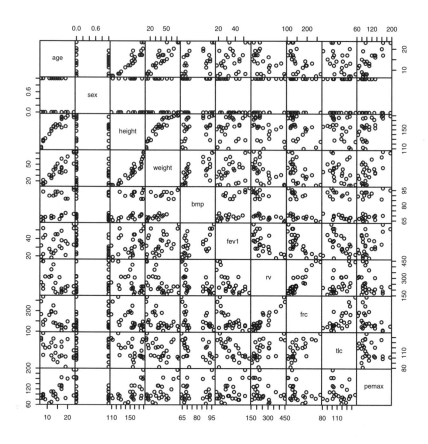

Figure 9.1. Pairwise plots for cystic fibrosis data.

You can obtain pairwise scatterplots between all the variables in the data set. This is done using the function `pairs`. To get Figure 9.1, you simply write

```
> par(mex=0.5)
> pairs(cystfibr, gap=0, cex.labels=0.9)
```

The arguments `gap` and `cex.labels` control the visual appearance by removing the space between subplots and decreasing the font size. The `mex` graphics parameter reduces the interline distance in the margins.

A similar plot is obtained by simply saying `plot(cystfibr)` since the `plot` function is generic and behaves differently depending on the class of its arguments (see Section 1.4.2). Here the argument is a data frame and a `pairs` plot is a fairly reasonable thing to get when asking for a plot of an entire data frame (although you might equally reasonably have expected a histogram or a bar chart of each variable instead).

The individual plots do get rather small, probably not suitable for direct publication, but it is quite an effective way of obtaining an overview of multidimensional issues. For example, the close relations among age, height, and weight appear clearly on the plot.

In order to be able to refer directly to the variables in cystfibr, we add it to the search path:

```
> attach(cystfibr)
```

Because this data set contains common variable names like height and weight, it is a good idea to ensure that you do not have identically named variables in the workspace at this point. In particular, such names were used in the introductory session.

9.2   Model specification and output

Specification of a multiple regression analysis is done by setting up a model formula with + between the explanatory variables:

```
lm(pemax~age+sex+height+weight+bmp+fev1+rv+frc+tlc)
```

which is meant to be read as "pemax is described using a model that is additive in age, sex, and so forth." (pemax is the maximal expiratory pressure. See Appendix B for a description of the other variables in cystfibr.)

As usual, there is not much output from lm itself, but with the aid of summary you can obtain some more interesting output:

```
> summary(lm(pemax~age+sex+height+weight+bmp+fev1+rv+frc+tlc))

Call:
lm(formula = pemax ~ age + sex + height + weight + bmp + fev1 +
    rv + frc + tlc)

Residuals:
    Min      1Q  Median      3Q     Max
-37.338 -11.532   1.081  13.386  33.405

Coefficients:
            Estimate Std. Error t value Pr(>|t|)
(Intercept) 176.0582   225.8912   0.779    0.448
age          -2.5420     4.8017  -0.529    0.604
```

```
sex          -3.7368    15.4598   -0.242    0.812
height       -0.4463     0.9034   -0.494    0.628
weight        2.9928     2.0080    1.490    0.157
bmp          -1.7449     1.1552   -1.510    0.152
fev1          1.0807     1.0809    1.000    0.333
rv            0.1970     0.1962    1.004    0.331
frc          -0.3084     0.4924   -0.626    0.540
tlc           0.1886     0.4997    0.377    0.711

Residual standard error: 25.47 on 15 degrees of freedom
Multiple R-Squared: 0.6373,      Adjusted R-squared: 0.4197
F-statistic: 2.929 on 9 and 15 DF, p-value: 0.03195
```

The layout should be wellknown by now. Notice that there is not one single significant t value, but the joint F test is nevertheless significant, so there must be an effect somewhere. The reason is that the t tests say anything about what happens only if you remove one variable and leave in all the others. You cannot see whether a variable would be statistically significant in a reduced model; all you can see is that no variable *must* be included.

Note further that there is quite a large difference between the unadjusted and the adjusted R^2, which is due to the large number of variables relative to the number of degrees of freedom for the variance. Recall that the former is the change in residual sum of squares relative to an empty model, whereas the latter is the similar change in residual *variance*:

```
> 1-25.5^2/var(pemax)
[1] 0.4183949
```

The 25.5 comes from "residual standard error" in the summary output.

The ANOVA table for a multiple regression analysis is obtained using anova and gives a rather different picture:

```
> anova(lm(pemax~age+sex+height+weight+bmp+fev1+rv+frc+tlc))
Analysis of Variance Table
```

```
Response: pemax
          Df  Sum Sq Mean Sq F value    Pr(>F)
age        1 10098.5 10098.5 15.5661  0.001296 **
sex        1   955.4   955.4  1.4727  0.243680
height     1   155.0   155.0  0.2389  0.632089
weight     1   632.3   632.3  0.9747  0.339170
bmp        1  2862.2  2862.2  4.4119  0.053010 .
fev1       1  1549.1  1549.1  2.3878  0.143120
rv         1   561.9   561.9  0.8662  0.366757
frc        1   194.6   194.6  0.2999  0.592007
tlc        1    92.4    92.4  0.1424  0.711160
Residuals 15  9731.2   648.7
---
```

```
Signif. codes:  0 `***' 0.001 `**' 0.01 `*' 0.05 `.' 0.1 ` ' 1
```

Note that except for the very last line ("tlc"), there is practically no correspondence between these *F* tests and the *t* tests from summary. In particular, the effect of age is now significant. That is because these tests are successive; they correspond to (reading upward from the bottom) a stepwise removal of terms from the model until finally only age is left. During the process, bmp came close to the magical 5% limit, but in view of the number of tests, this is hardly noteworthy.

The probability that 1 out of 8 independent tests gives a *p* value of 0.053 or below is actually just over 35%! The tests in the ANOVA table are not completely independent, but the approximation should be good.

The ANOVA table indicates that there is no significant improvements of the model once the age is included. It is possible to perform a joint test for whether *all* the other variables can be removed by adding up the sums of squares contributions and use the sum for an *F* test, that is:

```
> 955.4+155.0+632.3+2862.2+1549.1+561.9+194.6+92.4
[1] 7002.9
> 7002.9/8
[1] 875.3625
> 875.36/648.7
[1] 1.349407
> 1-pf(1.349407,8,15)
[1] 0.2935148
```

This corresponds to collapsing the eight lines of the table so that it would look like this:

```
            Df    Sum Sq  Mean Sq        F    Pr(>F)
age          1   10098.5  10098.5   15.566   0.00130
others       8    7002.9    875.4    1.349   0.29351
Residual    15    9731.2    648.7
```

(note that this is "cheat output" in which we have manually inserted the numbers computed above).

A procedure leading directly to the result is:

```
> m1<-lm(pemax~age+sex+height+weight+bmp+fev1+rv+frc+tlc)
> m2<-lm(pemax~age)
> anova(m1,m2)
Analysis of Variance Table

Model 1: pemax ~ age + sex + height + weight + bmp + fev1 + rv +
    frc + tlc
Model 2: pemax ~ age
  Res.Df      RSS Df Sum of Sq      F Pr(>F)
```

```
1     15  9731.2
2     23 16734.2 -8   -7002.9 1.3493 0.2936
```

which gives the appropriate *F* test with no manual computation.

Notice, however, that you need to be careful to ensure that the two models are actually nested. R does not check this although it does verify that the number of response observations is the same to safeguard against the more obvious mistakes. (When there are missing values in the descriptive variables, it's easy for the smaller model to contain more data points.)

From the ANOVA table we can thus see that it is allowable to remove all variables except age. However, that this particular variable is left in the model is primarily due to the fact that it was mentioned first in the model specification, as we see below.

9.3   Model search

R has the step() function for performing model searches by the Akaike Information Criterion. Since that is well beyond the scope of this book, we use simple manual variants of backward elimination.

In the following, we go through a practical model reduction for the example data. Notice that the output has been slightly edited to take up less space.

```
> summary(lm(pemax~age+sex+height+weight+bmp+fev1+rv+frc+tlc))
...
            Estimate Std. Error t value Pr(>|t|)
(Intercept) 176.0582   225.8912   0.779    0.448
age          -2.5420     4.8017  -0.529    0.604
sex          -3.7368    15.4598  -0.242    0.812
height       -0.4463     0.9034  -0.494    0.628
weight        2.9928     2.0080   1.490    0.157
bmp          -1.7449     1.1552  -1.510    0.152
fev1          1.0807     1.0809   1.000    0.333
rv            0.1970     0.1962   1.004    0.331
frc          -0.3084     0.4924  -0.626    0.540
tlc           0.1886     0.4997   0.377    0.711
...
```

One advantage of doing model reductions by hand is that you may impose some logical structure on the process. In the present case, it may, for instance, be natural to try to remove other lung function indicators first.

```
> summary(lm(pemax~age+sex+height+weight+bmp+fev1+rv+frc))
...
```

```
             Estimate Std. Error t value Pr(>|t|)
(Intercept) 221.8055   185.4350   1.196   0.2491
age          -3.1346     4.4144  -0.710   0.4879
sex          -4.6933    14.8363  -0.316   0.7558
height       -0.5428     0.8428  -0.644   0.5286
weight        3.3157     1.7672   1.876   0.0790 .
bmp          -1.9403     1.0047  -1.931   0.0714 .
fev1          1.0183     1.0392   0.980   0.3417
rv            0.1857     0.1887   0.984   0.3396
frc          -0.2605     0.4628  -0.563   0.5813
...
> summary(lm(pemax~age+sex+height+weight+bmp+fev1+rv))
...
             Estimate Std. Error t value Pr(>|t|)
(Intercept) 166.71822  154.31294   1.080   0.2951
age          -1.81783    3.66773  -0.496   0.6265
sex           0.10239   11.89990   0.009   0.9932
height       -0.40981    0.79257  -0.517   0.6118
weight        2.87386    1.55120   1.853   0.0814 .
bmp          -1.94971    0.98415  -1.981   0.0640 .
fev1          1.41526    0.74788   1.892   0.0756 .
rv            0.09567    0.09798   0.976   0.3425
...
> summary(lm(pemax~age+sex+height+weight+bmp+fev1))
...
             Estimate Std. Error t value Pr(>|t|)
(Intercept) 260.6313   120.5215   2.163   0.0443 *
age          -2.9062     3.4898  -0.833   0.4159
sex          -1.2115    11.8083  -0.103   0.9194
height       -0.6067     0.7655  -0.793   0.4384
weight        3.3463     1.4719   2.273   0.0355 *
bmp          -2.3042     0.9136  -2.522   0.0213 *
fev1          1.0274     0.6329   1.623   0.1219
...
> summary(lm(pemax~age+sex+height+weight+bmp))
...
             Estimate Std. Error t value Pr(>|t|)
(Intercept) 280.4482   124.9556   2.244   0.0369 *
age          -3.0750     3.6352  -0.846   0.4081
sex         -11.5281    10.3720  -1.111   0.2802
height       -0.6853     0.7962  -0.861   0.4001
weight        3.5546     1.5281   2.326   0.0312 *
bmp          -1.9613     0.9263  -2.117   0.0476 *
...
```

As seen, there was no obstacle to removing the four lung function variables. Next we try to reduce among the variables that describe the patient's state of physical development or size. Initially, we avoid removing weight and bmp, since they appear to be close to the 5% significance limit.

```
> summary(lm(pemax~age+height+weight+bmp))
```

```
...
            Estimate Std. Error t value Pr(>|t|)
(Intercept) 274.5307   125.5745   2.186   0.0409 *
age          -3.0832     3.6566  -0.843   0.4091
height       -0.6985     0.8008  -0.872   0.3934
weight        3.6338     1.5354   2.367   0.0282 *
bmp          -1.9621     0.9317  -2.106   0.0480 *
...
> summary(lm(pemax~height+weight+bmp))
...
            Estimate Std. Error t value Pr(>|t|)
(Intercept) 245.3936   119.8927   2.047   0.0534 .
height       -0.8264     0.7808  -1.058   0.3019
weight        2.7717     1.1377   2.436   0.0238 *
bmp          -1.4876     0.7375  -2.017   0.0566 .
...
> summary(lm(pemax~weight+bmp))
...
            Estimate Std. Error t value Pr(>|t|)
(Intercept) 124.8297    37.4786   3.331 0.003033 **
weight        1.6403     0.3900   4.206 0.000365 ***
bmp          -1.0054     0.5814  -1.729 0.097797 .
...
> summary(lm(pemax~weight))
...
            Estimate Std. Error t value Pr(>|t|)
(Intercept)  63.5456    12.7016   5.003 4.63e-05 ***
weight        1.1867     0.3009   3.944 0.000646 ***
...
```

Notice that once age and height were removed, bmp was no longer significant. In the original reference Altman (1991), weight, fev1, and bmp all ended up with p-values below 5%. However, far from all elimination procedures lead to that result.

It is also a good idea to pay close attention to the age, weight, and height variables, which are heavily correlated since we are dealing with children and adolescents.

```
> summary(lm(pemax~age+weight+height))
...
            Estimate Std. Error t value Pr(>|t|)
(Intercept) 64.65555   82.40935   0.785    0.441
age          1.56755    3.14363   0.499    0.623
weight       0.86949    0.85922   1.012    0.323
height      -0.07608    0.80278  -0.095    0.925
...
> summary(lm(pemax~age+height))
...
            Estimate Std. Error t value Pr(>|t|)
(Intercept) 17.8600    68.2493    0.262    0.796
age          2.7178     2.9325    0.927    0.364
```

```
height          0.3397     0.6900    0.492     0.627
...
> summary(lm(pemax~age))
...
            Estimate Std. Error t value Pr(>|t|)
(Intercept)    50.408     16.657    3.026  0.00601 **
age             4.055      1.088    3.726  0.00111 **
...
> summary(lm(pemax~height))
...
            Estimate Std. Error t value Pr(>|t|)
(Intercept) -33.2757    40.0445   -0.831  0.41453
height        0.9319     0.2596    3.590  0.00155 **
...
```

As it turns out, there is really no reason to prefer one of the three variables to the other two. The fact that an elimination method ends up with a model containing only weight is essentially a coincidence. You can easily be misled by model search procedures ending up with one highly significant variable — it is far from certain that the same variable would be chosen if you were to repeat the analysis on a new, similar data set.

What you may reasonably conclude is that there is probably a connection with the patient's physical development or size, which may be described in terms of age, height, or weight. Which description to use is arbitrary. If you want to choose one before the others, a decision cannot be based on the data, although possibly on theoretical considerations and/or results from previous investigations.

9.4   Exercises

9.1 The secher data are best analyzed after log-transforming birth weight as well as the abdominal and biparietal diameters. Fit a prediction equation for birth weight. How much is gained by using both diameters in a prediction equation? The sum of the two regression coefficients is almost exactly 3 — can this be given a nice interpretation?

9.2 The tlc data set contains a variable also called tlc. This may create some difficulties when analyzing it. Explain why, and suggest ways to overcome the problem. Describe tlc using the other variables in the data set and discuss the validity of the model.

9.3 The analyses of cystfibr involve sex, which is a binary variable. How would you interpret the results for this variable?

9.4 Consider the juul2 data set and select the group of those over 25 years old. Perform a regression analysis of $\sqrt{\text{igf1}}$ on age, and extend the model by including `height` and `weight`. Generate the analysis of variance table for the extended model. What is the surprise and why does it happen?

9.5 Analyze and interpret the effect of explanatory variables on the milk intake in the `kfm` data set using a multiple regression model. Notice that `sex` is a factor here; what does that imply for the analyses?

10

Linear models

Many data sets are inherently too complex to be handled adequately by standard procedures and thus require the formulation of ad-hoc models. The class of *linear models* provides a flexible framework into which many — although not all — of these cases can be fitted.

You may have noticed that the lm function is applied to data classified into groups (Chapter 6) as well as to (multiple) linear regression (Chapters 5 and 9) problems, even though the theory for these procedures appears to be quite different. However, they are, in fact, special cases of the same general model.

The basic point is that a multiple regression model can describe a wide variety of situations if you choose the explanatory variables suitably. There is no requirement that the explanatory variables should follow a normal distribution, or any continuous distribution for that matter. One simple example (which we use without comment in Chapter 9) is that a grouping into two categories can be coded as a 0/1 variable and used in a regression analysis. The regression coefficient in that case corresponds to a difference between two groups rather than the slope of an actual line. To encode a grouping with more than two categories, you can use multiple 0/1 variables.

Generating these *dummy variables* becomes tedious, but it can be automated by the use of model formulas. Among other things, they provide a convenient abstraction by treating classification variables (factors) and continuous variables symmetrically. You will need to learn exactly what model formulas do in order to become able to express your own modeling ideas.

This chapter contains a collection of models and their handling by `lm`, mainly in the form of relatively minor extensions and modifications of methods described earlier. It is meant only to give you a feel for the scope of possibilities and does not pretend to be complete.

10.1   Polynomial regression

One basic observation showing that multiple regression analysis can do more than meets the eye is that you can include second-order and higher powers of a variable in the model along with the original linear term. That is, you can have a model like

$$y = \alpha + \beta_1 x + \beta_2 x^2 + \cdots + \beta_k x^k + \epsilon$$

This obviously describes a nonlinear relation between y and x, but that does not matter; the model is still a linear model. What does matter is that the relation between the *parameters* and the expected observations is linear. It also does not matter that there is a deterministic relation between the regression variables $x, x^2, x^3, \ldots$ as long as there is no *linear* relation between them. However, fitting high-degree polynomials can be difficult because near-collinearity between terms makes the fit numerically unstable.

We return to the cystic fibrosis data set for an example. The plot of `pemax` and `height` in Figure 9.1 may suggest that the relation is not quite linear. One way to test this is to try to add a term that is the square of the height.

```
> data(cystfibr)
> attach(cystfibr)
> summary(lm(pemax~height+I(height^2)))
...
             Estimate Std. Error t value Pr(>|t|)
(Intercept) 615.36248  240.95580   2.554   0.0181 *
height        -8.08324    3.32052  -2.434   0.0235 *
I(height^2)    0.03064    0.01126   2.721   0.0125 *
...
```

Notice that the computed height^2 in the model formula needs to be "protected" by `I( ... )`. This technique is often used to prevent special interpretation of operators in a model formula. Such interpretation will not take place inside a function call, and `I` is the *identity* function that returns its argument unaltered.

We find a significant deviation from linearity. However, considering the process that led to doing this particular analysis, the p values have to be taken with more than a grain of salt. This is getting dangerously close to "data dredging", fishing expeditions in data. Consider it more an illustration of a technique than an exemplary data analysis.

To draw a plot of the fitted curve with prediction and confidence bands, we can use `predict`. To avoid problems caused by data not being sorted by height, we use `newdata`, which allows the prediction of values for a chosen set of predictors. Here we choose a set of heights between 110 and 180 cm in steps of 2 cm:

```
> pred.frame <- data.frame(height=seq(110,180,2))
> lm.pemax.hq <- lm(pemax~height+I(height^2))
> predict(lm.pemax.hq,interval="pred",newdata=pred.frame)
          fit      lwr      upr
1    96.90026 37.94461 155.8559
2    94.33611 36.82985 151.8424
3    92.01705 35.73077 148.3033
...
34 141.68922 88.70229 194.6761
35 147.21294 93.51117 200.9147
36 152.98174 98.36718 207.5963
```

Based on these predicted data, Figure 10.1 is obtained as follows:

```
> pp <- predict(lm.pemax.hq,newdata=pred.frame,interval="pred")
> pc <- predict(lm.pemax.hq,newdata=pred.frame,interval="conf")
> plot(height,pemax,ylim=c(0,200))
> matlines(pred.frame$height,pp,lty=c(1,2,2),col="black")
> matlines(pred.frame$height,pc,lty=c(1,3,3),col="black")
```

It is seen that the fitted curve is slightly decreasing for small heights. This is probably an artifact caused by the choice of a second-order polynomial to fit data. More likely, the reality is that `pemax` is relatively constant up to about 150 cm, whereafter it increases quickly with height. Note also that there seems to be a discrepancy between the prediction limits and the actual distribution of data for the smaller heights. The standard deviation might be larger for larger heights, but it is not impossible to obtain a similar distribution of points by coincidence. It is really not advisable to construct prediction intervals based on data as limited as these unless you are sure that the model is correct.

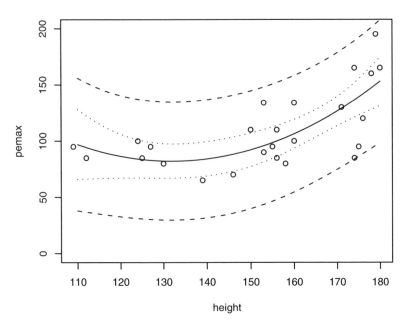

Figure 10.1. Quadratic regression with confidence- and prediction limits.

10.2   Regression through the origin

It sometimes makes sense to assume that a regression line passes through $(0, 0)$ — that the intercept of the regression line is zero. This can be specified in the model formula by adding the term `-1` ("minus intercept") to the right-hand side: `y ~ x - 1`.

The logic of the notation can be seen by writing the linear regression model as $y = \alpha \times 1 + \beta \times x + \epsilon$. The intercept corresponds to having an extra descriptive variable, which is the constant 1. Removing this variable yields regression through the origin.

This is a simulated example of a linear relationship through the origin $(y = 2x + \epsilon)$:

```
> x <- runif(20)
> y <- 2*x+rnorm(20,0,0.3)
> summary(lm(y~x))

Call:
lm(formula = y ~ x)
```

```
Residuals:
     Min       1Q    Median        3Q       Max
-0.50769  -0.08766   0.03802   0.14512   0.26358

Coefficients:
            Estimate Std. Error t value Pr(>|t|)
(Intercept) -0.14896    0.08812   -1.69    0.108
x            2.39772    0.15420   15.55 7.05e-12 ***
---
Signif. codes:  0 '***' 0.001 '**' 0.01 '*' 0.05 '.' 0.1 ' ' 1

Residual standard error: 0.2115 on 18 degrees of freedom
Multiple R-Squared: 0.9307,   Adjusted R-squared: 0.9269
F-statistic: 241.8 on 1 and 18 DF,  p-value: 7.047e-12

> summary(lm(y~x-1))

Call:
lm(formula = y ~ x - 1)

Residuals:
     Min       1Q    Median        3Q       Max
-0.62178  -0.16855  -0.04019   0.12044   0.27346

Coefficients:
  Estimate Std. Error t value Pr(>|t|)
x  2.17778    0.08669   25.12 4.87e-16 ***
---
Signif. codes:  0 '***' 0.001 '**' 0.01 '*' 0.05 '.' 0.1 ' ' 1

Residual standard error: 0.2216 on 19 degrees of freedom
Multiple R-Squared: 0.9708,   Adjusted R-squared: 0.9692
F-statistic: 631.1 on 1 and 19 DF,  p-value: 4.873e-16
```

In the first analysis, the intercept is not significant, which is, of course not surprising. In the second analysis we force the intercept to be zero, resulting in a slope estimate with a substantially improved accuracy.

Comparison of the R^2-values in the two analyses shows something that occasionally causes confusion: R^2 is much larger in the model with no intercept! This does *not*, however, mean that the relation is "more linear" when the intercept is not included or that more of the variation is explained. What is happening is that the definition of R^2 itself changes. It is most easily seen from the ANOVA tables in the two cases:

```
> anova(lm(y~x))
Analysis of Variance Table

Response: y
          Df Sum Sq Mean Sq F value    Pr(>F)
x          1 10.8134 10.8134  241.80 7.047e-12 ***
Residuals 18  0.8050  0.0447
```

```
---
Signif. codes:   0 '***' 0.001 '**' 0.01 '*' 0.05 '.' 0.1 ' ' 1
> anova(lm(y~x-1))
Analysis of Variance Table

Response: y
          Df   Sum Sq Mean Sq F value      Pr(>F)
x          1  30.9804 30.9804  631.06 4.873e-16 ***
Residuals 19   0.9328  0.0491
---
Signif. codes:   0 '***' 0.001 '**' 0.01 '*' 0.05 '.' 0.1 ' ' 1
```

Notice that the total sum of squares and the total number of degrees of freedom is not the same in the two analyses. In the model with an intercept there are 19 DF in all and the total sum of squares is $\sum(y_i - \bar{y})^2$, while the model without an intercept has a total of 20 DF and the total sum of squares is defined as $\sum y_i^2$. Unless $\bar{y}$ is close to zero, the latter "total SS" will be much larger than the former, so if the residual variance is similar, R^2 will be much closer to 1.

The reason for defining the total sum of squares like this for models without intercepts is that it has to correspond to the residual sum of squares in a minimal model. The minimal model has to be a submodel of the regression model; otherwise the ANOVA table simply does not make sense. In an ordinary regression analysis the minimal model is $y = \alpha + \epsilon$, but when the regression model does not include α, the only sensible minimal model is $y = 0 + \epsilon$.

10.3   Design matrices and dummy variables

The function model.matrix gives the *design matrix* for a given model. It can look like this:

```
> data(cystfibr)
> attach(cystfibr)
> model.matrix(pemax~height+weight)
   (Intercept) height weight
1            1    109   13.1
2            1    112   12.9
3            1    124   14.1
4            1    125   16.2
...
24           1    175   51.1
25           1    179   71.5
attr(,"assign")
[1] 0 1 2
```

You should not worry about the `"assign"` attribute at this stage, but the three columns are important. If you add them together, weighted by the corresponding regression coefficients, you get exactly the fitted values. Notice that the intercept enters as coefficient to a column of ones.

If the same is attempted for a model containing a factor, the following happens. We return to the anesthetic ventilation example on p. 113.

```
> data(red.cell.folate)
> attach(red.cell.folate)
> model.matrix(folate~ventilation)
     (Intercept) ventilationN2O+O2,op ventilationO2,24h
1            1                  0                 0
2            1                  0                 0
...
16           1                  1                 0
17           1                  1                 0
18           1                  0                 1
19           1                  0                 1
20           1                  0                 1
21           1                  0                 1
22           1                  0                 1
attr(,"assign")
[1] 0 1 1
attr(,"contrasts")
attr(,"contrasts")$ventilation
[1] "contr.treatment"
```

The two columns of zeros and ones are sometimes called *dummy variables*. They are interpreted exactly as above: Multiplying them by the respective regression coefficients and adding the results yields the fitted value. Notice that, for example, the second column is 1 for observations in group 2 and 0 otherwise; that is, the corresponding regression coefficient describes something that is added to the intercept for observations in that particular group. Both columns have zeros for observations from the first group, the mean value of which is described by the intercept (β_0) alone. The regression coefficient β_1 thus describes the *difference* in means between groups 1 and 2, and β_2 between groups 1 and 3.

You may be confused by the use of the term "regression coefficients" even though no regression lines are present in models like that above. The point is that you *formally* rewrite a model for groups as a multiple regression model, so that you can use the same software. As seen, there is a unique correspondence between the formal regression coefficients and the group means.

You can define dummy variables in several different ways to describe a grouping. This particular scheme is called *treatment contrasts* because if the first group is "no treatment" then the coefficients immediately give the treatment effects for each of the other groups. We do not discuss other choices here; see Venables and Ripley (2002) for a much deeper discussion. Note only that contrast type can be set on a per-term basis and that this is what is reflected in the "contrasts" attribute of the design matrix.

For completeness, this is what the "assign" attribute means: It indicates which columns belong together. When, for instance, you request an analysis of variance using anova, the sum of squares for ventilation will have 2 degrees of freedom, corresponding to removal of both columns simultaneously.

Removing the intercept from a model containing a factor term will not correspond to a model in which a particular group has mean zero, since such models are usually nonsensical. Instead, R generates a simpler set of dummy variables, which are indicator variables of the levels of the factor. This corresponds to the same model as when the intercept is included (the fitted values are identical), but the regression coefficients have a different interpretation.

10.4   Linearity over groups

Sometimes data are grouped according to a division of a continuous scale (e.g., by age group) or an experiment was designed to take several measurements at each of a fixed set of x-values. In both cases it is relevant to compare the results of a linear regression with those of an analysis of variance.

In the case of grouped x-values you might take a central value as representative for everyone in a given group, for instance formally letting everyone in a "20–29-year" category be 25 years old. If individual x-values are available, they may of course be used in a linear regression, but it makes the analysis a little more complicated so we discuss only the situation when that is not the case.

We thus have two alternative models for the same data. Both belong to the class of linear models that lm is capable of handling. The linear regression model is a *submodel* of the model for one-way analysis of variance, because the former can be obtained by placing restrictions on the parameters of the latter (namely that the true group means lie on a straight line).

It is possible to test whether or not a model reduction is allowable by comparing the reduction in the amount of variation explained to the residual variation in the larger model, resulting in an F test.

In the following example on trypsin concentrations in age groups (Altman, 1991, p. 212), data are given as the mean and SD within each of six groups. This is a kind of data that R is not quite prepared to handle, and it has therefore been necessary to create "fake" data giving the same means and SDs. These can be obtained via

```
> data(fake.trypsin)
> attach(fake.trypsin)
```

The actual results of the analysis of variance depend only on the means and SDs and are therefore independent of the faking. Readers interested in how to perform the actual faking should take a look at the file fake.trypsin.R in the data directory of the ISwR package.

The fake.trypsin data frame contains three variables, as seen by

```
> summary(fake.trypsin)
    trypsin             grp          grpf
 Min.   :-34.10   Min.   :1.000   1: 32
 1st Qu.:121.25   1st Qu.:2.000   2:137
 Median :165.47   Median :2.000   3: 38
 Mean   :168.68   Mean   :2.583   4: 44
 3rd Qu.:206.57   3rd Qu.:3.000   5: 16
 Max.   :357.95   Max.   :6.000   6:  4
```

Notice that there are both grp, which is a numerical vector, and grpf, which is a factor with six levels. Note also that the faking process involves random number generation in the data(fake.trypsin) step, so you will get varying results for the trypsin column.

Performing a one-way analysis of variance on the fake data gives the following ANOVA table:

```
> anova(lm(trypsin~grpf))
Analysis of Variance Table

Response: trypsin
           Df Sum Sq Mean Sq F value    Pr(>F)
grpf        5 224103   44821  13.508 9.592e-12 ***
Residuals 265 879272    3318
---
Signif. codes:  0 '***' 0.001 '**' 0.01 '*' 0.05 '.' 0.1 ' ' 1
```

If you had used grp instead of grpf in the model formula, you would have obtained a linear regression on the group number instead. In some circumstances that would have been a serious error, but here it actually makes sense. The midpoints of the age intervals are equidistant so the model is equivalent to assuming a linear development with age (the interpretation of the regression coefficient requires some care, though). The ANOVA table looks as follows:

```
> anova(lm(trypsin~grp))
Analysis of Variance Table

Response: trypsin
           Df Sum Sq Mean Sq F value    Pr(>F)
grp         1 206698  206698  62.009 8.451e-14 ***
Residuals 269 896677    3333
---
Signif. codes:  0 '***' 0.001 '**' 0.01 '*' 0.05 '.' 0.1 ' ' 1
```

Notice that the residual mean squares did not change very much, indicating that the two models describe the data nearly equally well. If you want to have a formal test of the simple linear model against the model where there is a separate mean for each group, it can be done easily as follows:

```
> model1 <- lm(trypsin~grp)
> model2 <- lm(trypsin~grpf)
> anova(model1,model2)
Analysis of Variance Table

Model 1: trypsin ~ grp
Model 2: trypsin ~ grpf
  Res.Df    RSS Df Sum of Sq      F Pr(>F)
1    269 896677
2    265 879272  4     17405 1.3114 0.2661
```

So we see that the model reduction has a nonsignificant p-value and hence that model2 does not fit data significantly better than model1.

This technique works *only* when one model is a submodel of the other, which is the case here since the linear model is defined by a restriction on the group means:

Another way to achieve the same result is formally to add the two models together as follows.

```
> anova(lm(trypsin~grp+grpf))
Analysis of Variance Table

Response: trypsin
           Df Sum Sq Mean Sq F value    Pr(>F)
grp         1 206698  206698 62.2959 7.833e-14 ***
grpf        4  17405    4351  1.3114    0.2661
Residuals 265 879272    3318
---
Signif. codes:  0 '***' 0.001 '**' 0.01 '*' 0.05 '.' 0.1 ' ' 1
```

This model is exactly the same as when only grpf was included. However, the ANOVA table now contains a subdivision of the model sum of squares in which the grpf line describes the *change* incurred by expand-

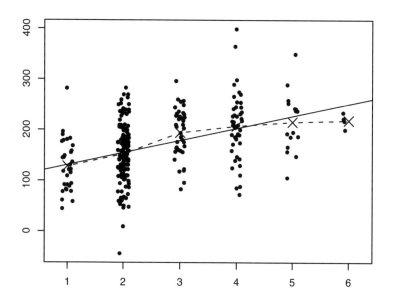

Figure 10.2. "Fake" data for the trypsin example with fitted line and empirical means.

ing the model from one to five parameters. The ANOVA table in Altman (1991, p. 213) is different, erroneously.

The plot in Figure 10.2 is made like this:

```
> xbar.trypsin <- tapply(trypsin,grpf,mean)
> stripchart(trypsin~grp,"jitter",jitter=.1,vertical=T,pch=20)
> lines(1:6,xbar.trypsin,type="b",pch=4,cex=2,lty=2)
> abline(lm(trypsin~grp))
```

The graphical techniques used here are essentially identical to those used for Figure 6.1, so do not go into further details.

Notice that the fakeness of the data is exposed by a point showing a negative trypsin concentration! The original data are unavailable but would likely show a distribution slightly skewed upward.

Actually, it *is* possible to analyze the data in R without generating fake data. A weighted regression analysis of the group means, with weights equal to the number of observations in each group, will yield the first two lines of the ANOVA table and the last one can be computed from the SDs. The details are as follows:

```
> n <- c(32,137, 38,44,16,4)
> tryp.mean <- c(128,152,194,207,215,218)
> tryp.sd <-c(50.9,58.5,49.3,66.3,60,14)
> gr<-1:6
> anova(lm(tryp.mean~gr+factor(gr),weights=n))
Analysis of Variance Table

Response: tryp.mean
           Df Sum Sq Mean Sq F value Pr(>F)
gr          1 206698  206698
factor(gr)  4  17405    4351
Residuals   0      0
```

Notice that the "Residuals" line is zero and that the F tests are not calculated. Omitting the factor(gr) term will cause that line to go into the Residuals and be treated as an estimate of the error variation, but that is not what you want since it does not include the information about the variation within groups. Instead, you need to fill in the missing information computed from the group standard deviations and sizes. The following gives the residual sum of squares and the corresponding degrees of freedom and mean squares:

```
> sum(tryp.sd^2*(n-1))
[1] 879271.9
> sum(n-1)
[1] 265
> sum(tryp.sd^2*(n-1))/sum(n-1)
[1] 3318.007
```

There is no simple way of updating the ANOVA table with an external variance estimate, but it is easy enough to do the computations directly:

```
> 206698/3318.007 # F statistic for gr
[1] 62.29583
> 1-pf(206698/3318.007,1,265) # p-value
[1] 7.838175e-14
> 4351/3318.007   # F statistic for factor(gr)
[1] 1.311329
> 1-pf(4351/3318.007,4,265) # p-value
[1] 0.2660733
```

10.5 Interactions

A basic assumption in a multiple regression model is that terms act additively on the response. However, this does not mean that linear models cannot describe nonadditivity. You can add special *interaction terms* that specify that the effect of one term is modified according to the level of another. In the model formulas in R such terms are generated using the colon operator, for example, a : b. Usually, you will also include the terms a and b, and R allows the notation a*b for a+b+a : b. Higher-order interactions among three or more variables are also possible.

The exact definition of the interaction terms and the interpretation of their associated regression coefficients can be elusive. Some peculiar things happen if an interaction term is present but one or more of the main effects is missing. The full details are probably best revealed through experimentation. However, depending on the nature of the terms a and b as factors or numerical variables, the overall effect of including interaction terms can be described as follows:

- *Interaction between two factors.* This is conceptually the simplest case. The model with interaction corresponds to having different levels for all possible combinations of levels of the two factors.

- *Interaction between a factor and a numerical variable.* In this case the model with interaction contains linear effects of the continuous variable, but with different slopes within each group defined by the factor.

- *Interaction between two continuous variables.* This gives a slightly peculiar model containing a new regression variable that is the product of the two. The interpretation is that you have a linear effect of varying one variable while keeping the other constant, but with a slope that changes as you vary the other variable.

10.6   Two-way ANOVA with replication

The coking data set comes from Johnson (1994, Section 13.1). The time to make coke from coal is analyzed in a 2×3 experiment varying the oven temperature and the oven width. There were three replications at each combination.

```
> data(coking)
> attach(coking)
> anova(lm(time~width*temp))
Analysis of Variance Table

Response: time
            Df  Sum Sq Mean Sq F value     Pr(>F)
width        2 123.143  61.572 222.102 3.312e-10 ***
temp         1  17.209  17.209  62.076 4.394e-06 ***
width:temp   2   5.701   2.851  10.283  0.002504 **
Residuals   12   3.327   0.277
```

```
---
Signif. codes:  0 `***' 0.001 `**' 0.01 `*' 0.05 `.' 0.1 ` ' 1
```

We see that the interaction term is significant. If we take a look at the cell means, we can get an idea of why this happens:

```
> tapply(time,list(width,temp),mean)
         1600       1900
4    3.066667 2.300000
8    7.166667 5.533333
12  10.800000 7.333333
```

The difference between high and low temperatures increases with oven width, making an additive model inadequate. When this is the case, the individual tests for the two factors make no sense. If the interaction had not been significant, then we would have been able to perform separate F tests for the two factors.

10.7 Analysis of covariance

As the example in this section we use a data set concerning growth conditions of *Tetrahymena* cells, collected by Per Hellung-Larsen. Data are from two groups of cell cultures where glucose was either added or not added to the growth medium. For each culture the average cell diameter (μ) and cell concentration (count per ml) were recorded. The cell concentration was set at the beginning of the experiment, and there is no systematic difference in cell concentration between the two glucose groups. However, it is expected that the cell diameter is affected by the presence of glucose in the medium.

Data are in the data frame hellung, which can be loaded and viewed like this:

```
> data(hellung)
> hellung
   glucose    conc diameter
1        1  631000     21.2
2        1  592000     21.5
3        1  563000     21.3
4        1  475000     21.0
...
49       2   14000     24.4
50       2   13000     24.3
51       2   11000     24.2
```

The coding of glucose is such that 1/2 means yes/no. There are no missing values.

Summarizing the data frame yields

```
> summary(hellung)
     glucose              conc              diameter
 Min.   :1.000   Min.   :  11000   Min.   :19.20
 1st Qu.:1.000   1st Qu.:  27500   1st Qu.:21.40
 Median :1.000   Median :  69000   Median :23.30
 Mean   :1.373   Mean   : 164325   Mean   :23.00
 3rd Qu.:2.000   3rd Qu.: 243000   3rd Qu.:24.35
 Max.   :2.000   Max.   : 631000   Max.   :26.30
```

Notice that the distribution of the concentrations is strongly right-skewed with a mean more than twice as big as the median. Note also that glucose is regarded as a numerical variable by summary, even though it has only two different values.

It will be more convenient to have glucose as a factor, so it is recoded as shown below. Recall that to change a variable inside a data frame, you use $-notation (p. 19) to specify the component you want to change:

```
> hellung$glucose <- factor(hellung$glucose, labels=c("Yes","No"))
> summary(hellung)
 glucose          conc              diameter
 Yes:32   Min.   :  11000   Min.   :19.20
 No :19   1st Qu.:  27500   1st Qu.:21.40
          Median :  69000   Median :23.30
          Mean   : 164325   Mean   :23.00
          3rd Qu.: 243000   3rd Qu.:24.35
          Max.   : 631000   Max.   :26.30
```

It is convenient to be able to refer to the variables of hellung without the hellung$ prefix, so we put hellung in the search path.

```
> attach(hellung)
```

10.7.1 Graphical description

First we plot the raw data (Figure 10.3):

```
> plot(conc,diameter,pch=as.numeric(glucose))
```

By calculating as.numeric(glucose), we convert the factor glucose to the underlying codes, 1 and 2. The specification of pch thus implies that group 1 ("Yes") is drawn using plotting character 1 (circles) and group 2 with plotting character 2 (triangles).

To get different plotting symbols, you must first create a vector containing the symbol numbers and give that as the pch argument. The following

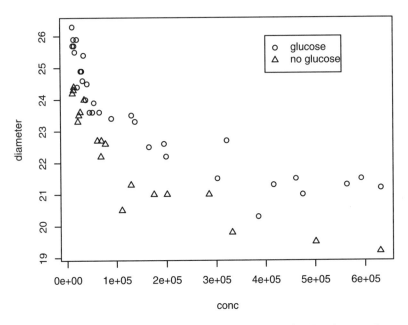

Figure 10.3. Plot of diameter versus concentration for *Tetrahymena* data.

form yields open and filled circles: `c(1,16)[glucose]`. It looks a bit cryptic at first, but it is really just a consequence of R's way of indexing. For indexing purposes, a factor like `glucose` behaves as a vector of 1s and 2s, so you get the first element of `c(1,16)`, namely 1, whenever an observation is from group 1; when the observation is from group 2 you similarly get 16.

The explanatory text is inserted with `legend` like this:

```
> legend(locator(n=1),legend=c("glucose","no glucose"),pch=1:2)
```

Notice that both the function and one of its arguments is named `legend`.

The function `locator` returns the coordinates of a point on a plot. It works so that the function awaits a click with a mouse button and then returns the cursor position. You may want to call `locator()` directly from the command line to see the effect. Notice that if you do not specify a value for n, then you need to right-click when you are done selecting points.

The plot shows a clear inverse and nonlinear relation between concentration and cell diameter. Further, it is seen that the cultures without glucose are systematically below cultures with added glucose.

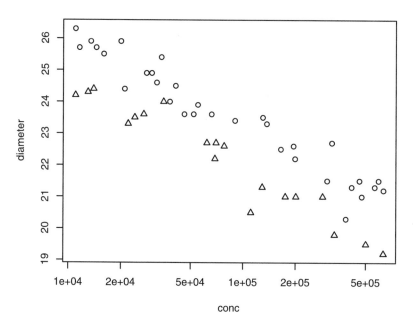

Figure 10.4. *Tetrahymena* data with logarithmic *x*-axis.

You get a much nicer plot (Figure 10.4) by using a logarithmic *x*-axis:

```
> plot(conc,diameter,pch=as.numeric(glucose),log="x")
```

Now the relation suddenly looks linear!

You could also try a log-log plot (shown in Figure 10.5 with regression lines as described below):

```
> plot(conc,diameter,pch=as.numeric(glucose),log="xy")
```

As seen, this really does not change much, but it was nevertheless decided to analyze data with both diameter and concentration log-transformed, because a power-law relation was expected ($y = \alpha x^\beta$, which gives a straight line on a log-log plot).

When adding regression lines to a log plot or log-log plot, you should notice that abline interprets them as lines in the coordinate system obtained *after* taking (base-10) logarithms. Thus, you can add a line for each group with abline applied to the results of a regression analysis of log10(diameter) on log10(conc). First, however, it is convenient to define data frames corresponding to the two glucose groups:

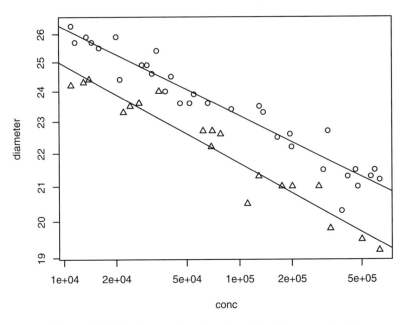

Figure 10.5. *Tetrahymena* data, log-log plot with regression lines.

```
> tethym.gluc <- hellung[glucose=="Yes",]
> tethym.nogluc <- hellung[glucose=="No",]
```

Notice that you have to use the names, not the numbers, of the factor levels.

Since we only need the two data frames for adding lines to the figure, it would be cumbersome to add them in turn to the search path with attach, do the plotting, and then use detach to remove them. It is easier to use the data argument to lm; this allows you to explicitly specify the data frame in which to look for variables. The two regression lines are drawn with

```
> lm.nogluc <- lm(log10(diameter)~ log10(conc),data=tethym.nogluc)
> lm.gluc <- lm(log10(diameter)~ log10(conc),data=tethym.gluc)
> abline(lm.nogluc)
> abline(lm.gluc)
```

— whereafter the plot looks like Figure 10.5. It is seen that the lines fit the data quite well and that they are almost, but not perfectly, parallel. The question is whether the difference in slope is statistically significant. This is the topic of the next section.

10.7.2   Comparison of regression lines

Corresponding to the two lines from before, we have the following regression analyses:

```
> summary(lm(log10(diameter)~ log10(conc), data=tethym.gluc))

Call:
lm(formula = log10(diameter) ~ log10(conc), data = tethym.gluc)

Residuals:
      Min         1Q      Median        3Q         Max
-0.0267219 -0.0043361  0.0006891  0.0035489  0.0176077

Coefficients:
             Estimate Std. Error t value Pr(>|t|)
(Intercept)  1.63134    0.01345  121.29   <2e-16 ***
log10(conc) -0.05320    0.00272  -19.56   <2e-16 ***
---
Signif. codes:  0 '***' 0.001 '**' 0.01 '*' 0.05 '.' 0.1 ' ' 1

Residual standard error: 0.008779 on 30 degrees of freedom
Multiple R-Squared: 0.9273,     Adjusted R-squared: 0.9248
F-statistic: 382.5 on 1 and 30 DF,   p-value: < 2.2e-16

> summary(lm(log10(diameter)~ log10(conc), data=tethym.nogluc))

Call:
lm(formula = log10(diameter) ~ log10(conc), data = tethym.nogluc)

Residuals:
      Min         1Q      Median        3Q         Max
-2.192e-02 -4.977e-03  5.598e-05  5.597e-03  1.663e-02

Coefficients:
             Estimate Std. Error t value Pr(>|t|)
(Intercept)  1.634761   0.020209   80.89  < 2e-16 ***
log10(conc) -0.059677   0.004125  -14.47 5.48e-11 ***
---
Signif. codes:  0 '***' 0.001 '**' 0.01 '*' 0.05 '.' 0.1 ' ' 1

Residual standard error: 0.009532 on 17 degrees of freedom
Multiple R-Squared: 0.9249,     Adjusted R-squared: 0.9205
F-statistic: 209.3 on 1 and 17 DF,   p-value: 5.482e-11
```

Notice that you can use arithmetic expressions in the model formula [here `log10(...)`]. There are limitations, though, because, for example, `z~x+y` means a model where z is described by an additive model in x and y, which is not the same as a regression analysis on the sum of the two. The latter may be specified using `z~I(x+y)` (`I` for "identity").

A quick assessment of the significance of the difference between the slopes of the two lines can be obtained as follows: The difference between the slope estimates is 0.0065, and the standard error of that is $\sqrt{0.0041^2 + 0.0027^2} = 0.0049$. Since $t = 0.0065/0.0049 = 1.3$, it would seem that we are allowed to assume that the slopes are the same.

It is, however, preferable to fit a model to the entire data set and test the hypothesis of equal slopes in that model. One reason that this approach is preferable is that it can be generalized to more complicated models. Another reason is that even though there is nothing seriously wrong with the simple test for equal slopes, that procedure gives you little information on how to proceed. If the slopes are the same, you would naturally want to find an estimate of the common slope and of the distance between the parallel lines.

First we set up a model that allows the relation between concentration and cell diameter to have different slopes and intercepts in the two glucose groups:

```
> summary(lm(log10(diameter)~log10(conc)*glucose))

Call:
lm(formula = log10(diameter) ~ log10(conc) * glucose)

Residuals:
      Min         1Q     Median         3Q        Max
-2.672e-02 -4.888e-03  5.598e-05  3.767e-03  1.761e-02

Coefficients:
                        Estimate Std. Error t value Pr(>|t|)
(Intercept)             1.631344   0.013879 117.543   <2e-16 ***
log10(conc)            -0.053196   0.002807 -18.954   <2e-16 ***
glucoseNo               0.003418   0.023695   0.144    0.886
log10(conc):glucoseNo  -0.006480   0.004821  -1.344    0.185
---
Signif. codes:  0 `***' 0.001 `**' 0.01 `*' 0.05 `.' 0.1 ` ' 1

Residual standard error: 0.009059 on 47 degrees of freedom
Multiple R-Squared: 0.9361,     Adjusted R-squared: 0.9321
F-statistic: 229.6 on 3 and 47 DF,  p-value: < 2.2e-16
```

These regression coefficients should be read as follows: The expected value of the log cell diameter for an observation with cell concentration C is obtained as the sum of the following four quantities:

1. The intercept, 1.6313

2. $-0.0532 \times \log_{10} C$

3. 0.0034, but only for a culture without glucose

4. $-0.0065 \times \log_{10} C$, only for cultures without glucose

Accordingly, for cell cultures with glucose, we have the linear relation

$$\log_{10} D = 1.6313 - 0.0532 \times \log_{10} C$$

and for cultures without glucose we have

$$\log_{10} D = (1.6313 + 0.0034) - (0.0532 + 0.0065) \times \log_{10} C$$

Put differently, the first two coefficients in the joint model can be interpreted as the estimates for intercept and slope in group 1, whereas the latter two are the differences between group 1 and group 2 in intercept and slope, respectively. Comparison with the separate regression analyses shows that slopes and intercepts are the same as in the joint analysis. The standard errors differ a little from the separate analyses because a pooled variance estimate is now used. Notice that the rough test of difference in slope outlined above is essentially the t test for the last coefficient.

Notice also that the glucose and log10(conc).glucose terms indicate items to be added for cultures *without* glucose. This is because the factor levels are ordered yes = 1 and no = 2, and the base level is the first group.

Fitting an additive model, we get

```
> summary(lm(log10(diameter)~log10(conc)+glucose))
...
Coefficients:
             Estimate Std. Error t value Pr(>|t|)
(Intercept)  1.642132   0.011417  143.83  < 2e-16 ***
log10(conc) -0.055393   0.002301  -24.07  < 2e-16 ***
glucoseNo   -0.028238   0.002647  -10.67 2.93e-14 ***
...
```

Here the interpretation of the coefficients is that the estimated relation for cultures with glucose is

$$\log_{10} D = 1.6421 - 0.0554 \times \log_{10} C$$

and for cultures without glucose it is

$$\log_{10} D = (1.6421 - 0.0282) - 0.0554 \times \log_{10} C$$

That is, the lines for the two cultures are parallel, but the log diameters for cultures without glucose are 0.0282 below those with glucose. On the original (nonlogarithmic) scale, this means that the former are 6.3% lower (a constant absolute difference on a logarithmic scale corresponds to constant relative differences on the original scale and $10^{-0.0282} = 0.937$).

The joint analysis presumes that the variance around the regression line is the same in the two groups. This assumption should really have been tested before embarking on the above analysis. A formal test can be performed with var.test, which conveniently allows a pair of linear models as arguments instead of a model formula or two group vectors:

```
> var.test(lm.gluc,lm.nogluc)

        F test to compare two variances

data:  lm.gluc and lm.nogluc
F = 0.8482, num df = 30, denom df = 17, p-value = 0.6731
alternative hypothesis: true ratio of variances is not equal to 1
95 percent confidence interval:
 0.3389901 1.9129940
sample estimates:
ratio of variances
          0.8481674
```

When there are more than two groups, Bartlett's test is the one to use. It too allows linear models to be compared. The reservations about robustness against nonnormality apply here, too.

It is seen that it is possible to assume that the lines have the same slope and that they have the same intercept, but — as we see below — not both at once. The hypothesis of a common intercept is silly anyway unless the slopes are also identical: The intercept is by definition the y-value at $x = 0$, which because of the log scale corresponds to a cell concentration of 1. That is far outside the region the data cover, and it is a completely arbitrary point, which will change if the concentrations are measured in different units.

The ANOVA table for the model is

```
> anova(lm(log10(diameter)~ log10(conc)*glucose))
Analysis of Variance Table

Response: log10(diameter)
                    Df   Sum Sq  Mean Sq F value    Pr(>F)
log10(conc)          1 0.046890 0.046890 571.436 < 2.2e-16 ***
glucose              1 0.009494 0.009494 115.698  2.89e-14 ***
log10(conc):glucose  1 0.000148 0.000148   1.807    0.1853
Residuals           47 0.003857 0.000082
---
Signif. codes:  0 '***' 0.001 '**' 0.01 '*' 0.05 '.' 0.1 ' ' 1
```

The model formula a*b, where in the present case a is log10(conc) and b is glucose, is a short form for a + b + a:b, which is read "effect of a plus effect of b plus interaction". The F test in the penultimate line of the ANOVA table is a test for the hypothesis that the last term (a:b) can be omitted, reducing the model to be additive in log10(conc) and glucose, which corresponds to the parallel regression lines. The F test one line earlier indicates whether you can *subsequently* remove glucose, and the one in the first line to removing log10(conc), leaving an empty model.

Alternatively, you can read the table from top to bottom as adding terms describing more and more of the total sum of squares. To those familiar with the SAS system, this kind of ANOVA table is known as type I sums of squares.

The *p* value for log10(conc):glucose can be recognized as that of the *t* test for the coefficient labeled log10(conc).glucose in the previous output. The *F* statistic is exactly the square of *t* as well. However, this is true only because there are just two groups. Had there been three or more then there would have been several regression coefficients and the *F* test would have tested them all against zero simultaneously, just like when all groups are tested equal in a one-way analysis of variance.

Note that the test for removing log10(conc) does not make sense, because you would have to remove glucose first, which is "forbidden" when glucose has a highly significant effect. It makes perfectly good sense to test log10(conc) *without* removing glucose — that corresponds to testing that the two parallel regression lines can be assumed horizontal — but that test is not found in the ANOVA table. You can get the right test by changing the order of terms on the model formula; compare, for instance, these two regression analyses:

```
> anova(lm(log10(diameter)~glucose+log10(conc)))
Analysis of Variance Table

Response: log10(diameter)
            Df   Sum Sq  Mean Sq F value   Pr(>F)
glucose      1 0.008033 0.008033  96.278 4.696e-13 ***
log10(conc)  1 0.048351 0.048351 579.494 < 2.2e-16 ***
Residuals   48 0.004005 0.000083
---
Signif. codes:  0 '***' 0.001 '**' 0.01 '*' 0.05 '.' 0.1 ' ' 1
> anova(lm(log10(diameter)~log10(conc)+ glucose))
Analysis of Variance Table

Response: log10(diameter)
            Df   Sum Sq  Mean Sq F value   Pr(>F)
log10(conc)  1 0.046890 0.046890  561.99 < 2.2e-16 ***
glucose      1 0.009494 0.009494  113.78 2.932e-14 ***
Residuals   48 0.004005 0.000083
---
Signif. codes:  0 '***' 0.001 '**' 0.01 '*' 0.05 '.' 0.1 ' ' 1
```

They both describe exactly the same model, as seen by the residual sum of squares being identical. The partitioning of the sum of squares is *not* the same, though — and the difference may be much more dramatic than it is here. The difference is whether log10 (conc) is added to a model already containing glucose, or vice versa. Since the second F test in both tables is highly significant, no model reduction is possible and the F test in the line above it is irrelevant.

If you go back and look at the regression coefficients in the model with parallel regression lines, you will see that the squares of the t tests are 579.49 and 113.8, precisely the last F test in the above two tables.

It is informative to compare the above covariance analysis with the simpler analysis in which the effect of cell concentration is ignored:

```
> t.test(log10(diameter)~glucose)

        Welch Two Sample t-test

data:  log10(diameter) by glucose
t = 2.7037, df = 36.31, p-value = 0.01037
alternative hypothesis: true difference in means is not equal to 0
95 percent confidence interval:
 0.006492194 0.045424241
sample estimates:
mean in group Yes  mean in group No
         1.370046          1.344088
```

Notice that the p-value is much less extreme. It is still significant in this case, but in smaller data sets the statistical significance could easily disappear completely. The difference in mean between the two groups is 0.026, which is comparable to the 0.028 that was the glucose effect in the analysis of covariance. However, the confidence interval goes from 0.006 to 0.045, where the analysis of covariance had 0.023 to 0.034 [$0.0282 \pm t_{.975}(48) \times 0.0026$], which is almost four times as narrow, obviously a substantial gain in efficiency.

10.8 Diagnostics

Regression diagnostics are used to evaluate the model assumptions and investigate whether or not there are observations with a large influence on the analysis. A basic set of these is available via the plot method for lm objects. Four different plots are in the set, so it is convenient to display them in a 2×2 layout (Figure 10.6):

```
> data(thuesen)
> attach(thuesen)
> options(na.action="na.exclude")
> lm.velo <- lm(short.velocity~blood.glucose)
> par(mfrow=c(2,2), mex=0.6)
```

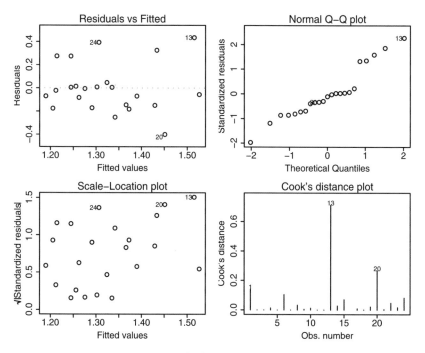

Figure 10.6. Default regression diagnostics.

```
> plot(lm.velo)
> par(mfrow=c(1,1), mex=1)
```

The par commands set up for a 2 × 2 layout with compressed margin texts and go back to normal after plotting.

The first panel shows residuals versus fitted values. The second is a Q–Q normal distribution plot of standardized residuals. Notice that there are residuals and standardized residuals; the latter have been corrected for differences in the SD of residuals depending on their position in the design. (Residuals corresponding to extreme x-values generally have a lower SD due to overfitting.) The third plot is of the square root of the absolute value of the standardized residuals; this reduces the skewness of the distribution and makes it much easier to detect if there might be a trend in the dispersion. The fourth plot is of "Cook's distance" which is a measure of the influence of each observation on the regression coefficients. We return to Cook's distance shortly.

The plots for the thuesen data show observation no. 13 as extreme in several respects. It has the largest residual as well as a prominent spike in the Cook's distance plot. Observation no. 20 also has a large residual, but not quite as conspicuous a Cook's distance.

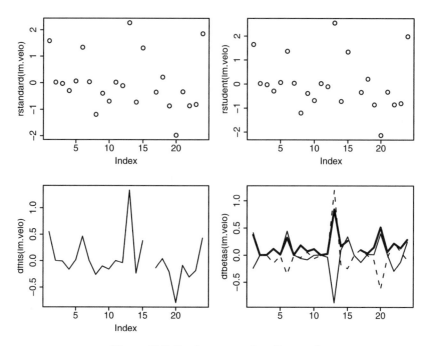

Figure 10.7. Further regression diagnostics.

```
> par(mfrow=c(2,2), mex=0.6)
> plot(rstandard(lm.velo))
> plot(rstudent(lm.velo))
> plot(dffits(lm.velo),type="l")
> matplot(dfbetas(lm.velo),type="l", col="black")
> lines(sqrt(cooks.distance(lm.velo)), lwd=2)
> par(mfrow=c(1,1), mex=1)
```

It is also possible to obtain individual diagnostics; a selection is shown in Figure 10.7. The function rstandard gives the standardized residuals discussed above. There is also rstudent, which gives *leave-out-one residuals*, in which the fitted value is calculated omitting the current point; if the model is correct, then these will follow a (Student's) *t* distribution. (Unfortunately, some texts use "studentized residuals" for residuals divided by their standard deviation, i.e., what rstandard calculates in R.) It takes a keen eye to see the difference between the two types of residuals, but the extreme residuals tend to be a little further out in the case of rstudent.

The function dffits expresses how much an observation affects the associated fitted value. As with the residuals, observations 13 and maybe 20 seem to stick out. Notice that there is a gap in the line. This is due to the missing observation 16 and the use of na.exclude. This looks a little awkward but has the advantage of making the x-axis match the observation number.

The function dfbetas gives the change in the estimated parameters if an observation is excluded, relative to its standard error. It is a matrix, so matplot is useful to plot them all in one plot. Notice that observation 13 affects both α (the solid line) and β by nearly one standard error.

The name dfbetas refers to its use in multiple regression analysis, where you write the model as $y = \beta_0 + \beta_1 x_1 + \beta_2 x_2 + \cdots$. This gets a little confusing in a simple regression analysis where the intercept is otherwise called α.

Cook's distance D calculated by cooks.distance is essentially a joint measure of the components of dfbetas. The exact procedure is to take the *unnormalized* change in coefficients and use the norm defined by the estimated covariance matrix for $\hat{\beta}$, and then divide by the number of coefficients. $\sqrt{D}$ is on the the the same scale as dfbetas and was added to that plot as a double-width line. (If you look inside the R functions for some of these quantities, you will find them apparently quite different from the descriptions above, but they are in fact the same, only computationally more efficient.)

Thus, the picture is that observation 13 seems to be influential. Let's look at the analysis without this observation:

We use the subset argument to lm, which, like other indexing operations, can be used with negative numbers to remove observations.

```
> summary(lm(short.velocity~blood.glucose, subset=-13))

Call:
lm(formula = short.velocity ~ blood.glucose, subset = -13)

Residuals:
     Min       1Q   Median       3Q      Max
-0.31346 -0.11136 -0.01247  0.06043  0.40794

Coefficients:
               Estimate Std. Error t value Pr(>|t|)
(Intercept)     1.18929    0.11061  10.752 9.22e-10 ***
blood.glucose   0.01082    0.01029   1.052    0.305
---
Signif. codes:  0 `***' 0.001 `**' 0.01 `*' 0.05 `.' 0.1 ` ' 1

Residual standard error: 0.193 on 20 degrees of freedom
Multiple R-Squared: 0.05241,	Adjusted R-squared: 0.005026
F-statistic: 1.106 on 1 and 20 DF,  p-value: 0.3055
```

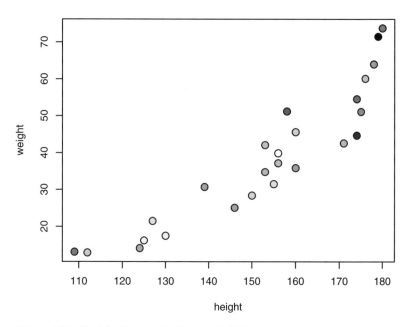

Figure 10.8. Cook's distance (colour coded) in pemax ~ height + weight.

The relation practically vanished in thin air! The whole analysis actually hinges on a single observation. If data and model are valid, then of course the original *p*-value is correct, and perhaps you could also say that there will always be influential observations in small datasets, but some caution in the interpretation does seem advisable.

The methods for finding influential observations and outliers are even more important in regression analysis with multiple descriptive variables. One of the big problems is how to present the quantities graphically in a sensible way. This might be done using three-dimensional plots (the add-on package scatterplot3d makes this possible), but you can get quite far using colour coding.

Here, we see how to display the value of Cook's distance (which is always positive) graphically for a model where pemax is described using height and weight:

```
> cookd <- cooks.distance(lm(pemax~height+weight))
> cookd <- cookd/max(cookd)
> cook.colors <- gray(1-sqrt(cookd))
> plot(height,weight,bg=cook.colors,pch=21,cex=1.5)
> points(height,weight,pch=1,cex=1.5)
```

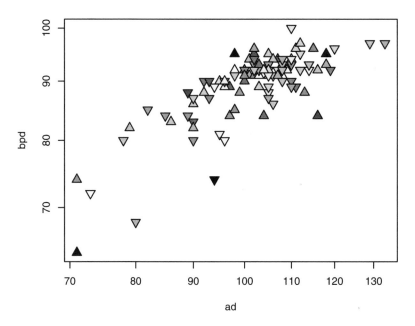

Figure 10.9. Studentized residuals in the Secher data, colour coded. Positive values are marked with upward-pointing triangles; negative ones point down.

The above is how Figure 10.8 is drawn. The first line computes Cook's distance and the second scales it to a value between 0 and 1. Thereafter, a colour coding of the values in cookd is made with the function gray. The latter interprets its argument as degree of whiteness, so if you want a large distance represented as black, you need to subtract the value from 1. Furthermore, it is convenient to take the square root of cookd because it is a quadratic distance measure (which in practice shows up in the form of too many white or nearly white points). Then a scatterplot of height versus weight is drawn with the chosen colors. A filled plotting symbol in enlarged symbol size is used to get the grayscale to stand out more clearly.

You can use similar techniques to describe other influence measures. In the case of signed measures, you might use different symbols for positive and negative values. Here is an example on Studentized residuals in a data set describing birth weight as a function of abdominal and biparietal diameters determined by ultrasonography of the fetus immediately before birth, also used in Exercise 9.1 (Figure 10.9):

```
> data(secher)
> attach(secher)
> rst <- rstudent(lm(log10(bwt)~log10(ad)+log10(bpd)))
```

```
> range(rst)
[1] -3.707509  3.674050
> rst <- rst/3.71
> plot(ad,bpd,log="xy",bg=gray(1-abs(rst)),
+      pch=ifelse(rst>0,24,25), cex=1.5)
```

10.9   Exercises

10.1 Set up an additive model for the `ashina` data (see Exercise 4.6), containing additive effects of subjects, period, and treatment. Compare the results with those obtained from t tests.

10.2 Perform a two-way analysis of variance on the `tb.dilute` data. Modify the model to have a dose effect that is linear in log dose. Compute a confidence interval for the slope. An alternative approach could be to calculate a slope for each animal and perform a test based on these. Compute a confidence interval for the mean slope, and compare it to the preceding result.

10.3 Consider the following definitions:

```
a <- gl(2, 2, 8)
b <- gl(2, 4, 8)
x <- 1:8
y <- c(1:4,8:5)
z <- rnorm(8)
```

Generate the model matrices for models z ~ a*b, z ~ a:b, etc. Discuss the implications. Carry out the model fits, and notice which models contain singularities.

10.4 (Advanced) In the `secretin` experiment you may expect to find interindividual differences, not only between the level of glucose, but also in the change induced by the injection of secretin. The factor `time.comb` combines time values at 30, 60, and 90 minutes. The factor `time20plus` combines all values from 20 minutes and onward. Discuss the differences and relations among the following linear models:

```
model1 <- lm(gluc ~ person * time)
model2 <- lm(gluc ~ person + time)
model3 <- lm(gluc ~ person * time20plus + time)
model4 <- lm(gluc ~ person * time20plus + time.comb)
```

10.5 Analyze the blood pressure in the `bp.obese` data set as a function of obesity and gender.

10.6 Analyze the `vitcap2` data set using analysis of covariance. Revisit Exercise 4.2 and compare the conclusions. Try using the `drop1` function with `test="F"` instead of `summary` in this model.

10.7   In the `juul` data set make regression analyses for prepubescent children (Tanner stage 1) of $\sqrt{\texttt{igf1}}$ versus age separately for boys and girls. Compare the two regression lines.

10.8   Try `step` on the `kfm` data and discuss the result. One observation appears to be influential on the diagnostic plot for this model — explain why. What happens if you reduce the model further?

10.9   For the `juul` data, fit a model for `igf1` with interactions between age, sex, and Tanner stage for those under 25 years old. Explain the interpretation of this model. Hint: A plot of the fitted values against age should be helpful. Use diagnostic plots to evaluate possible transformations of the dependent variable: untransformed, log, or square root.

11
Logistic regression

Sometimes you wish to model *binary outcomes*, variables that can have only two possible values: diseased/nondiseased, and so forth. For instance, you want to describe the risk of getting a disease depending on various kinds of exposures. Chapter 7 discusses some simple techniques based on tabulation, but you might also want to model dose-response relationships (where the predictor is a continuous variable) or model the effect of multiple variables simultaneously. It would be very attractive to be able to use the same modeling techniques as for linear models.

However, it is not really attractive to use additive models for probabilities since they have a limited range and regression models could predict off-scale values below zero or above 1. It makes better sense to model the probabilities on a transformed scale; this is what is done in logistic regression analysis.

A linear model for transformed probabilities can be set up as

$$\text{logit } p = \beta_0 + \beta_1 x_1 + \beta_2 x_2 + \ldots \beta_k x_k$$

in which $\text{logit } p = \log[p/(1-p)]$ is the *log odds*. A constant additive effect on the logit scale corresponds to a constant odds ratio. The choice of the logit function is not the only one possible, but it has some mathematically convenient properties. Other choices do exist; the probit function (the quantile function of the normal distribution) or $\log(-\log p)$, which has a connection to survival analysis models.

One thing to notice about the logistic model is that there is no error term as in linear models. We are modelling the probability of an event directly, and that in itself will determine the variability of the binary outcome. There is no variance parameter like in the normal distribution.

The parameters of the model can be estimated by the *method of maximum likelihood*. This is a quite general technique, similar to the least-squares method in that it finds a set of parameters that optimizes a goodness-of-fit criterion (in fact, the least-squares method itself is a slightly modified maximum-likelihood procedure). The *likelihood function* $L(\beta)$ is simply the probability of the entire observed data set for varying parameters.

The *deviance* is the difference between the maximized value of $-2 \log L$ and the similar quantity under a "maximal model" that fits data perfectly. Changes in deviance caused by a model reduction will be approximately χ^2-distributed with degrees of freedom equal to the change in the number of parameters.

In this chapter we see how to perform logistic regression analysis in R. There naturally is quite a large overlap with the material on linear models since the description of models is quite similar, but there are also some special issues concerning deviance tables and the specification of models for pretabulated data.

11.1   Generalized linear models

Logistic regression analysis belongs to the class of *generalized linear models*. These models are characterized by their response distribution (here, the binomial distribution) and a *link function*, which transfers the mean value to a scale in which the relation to background variables is described as linear/additive. In a logistic regression analysis the link function is logit $p = \log[p/(1-p)]$.

There are several other examples of generalized linear models; for instance, analysis of count data is often handled by the multiplicative Poisson model where the link function is $\log \lambda$, with λ the mean of the Poisson distributed observation. All of these models can be handled using the same algorithm, which also allows the user some freedom to define his or her own models by defining suitable link functions.

In R generalized linear models are handled by the `glm` function. This function is very similar to `lm`, which we have used many times for linear normal models. The two functions use essentially the same model formulas and extractor functions (`summary`, etc.), but `glm` also needs to have specified *which* generalized linear model is desired. This is done via the `family` argument. To specify a binomial model with logit link (i.e., logistic regression analysis), you write `family=binomial("logit")`.

11.2 Logistic regression on tabular data

In this section we analyze the example concerning hypertension from Altman (1991, p. 353). First, we need to enter data, which is done as follows:

```
> no.yes <- c("No","Yes")
> smoking <- gl(2,1,8,no.yes)
> obesity <- gl(2,2,8,no.yes)
> snoring <- gl(2,4,8,no.yes)
> n.tot <- c(60,17,8,2,187,85,51,23)
> n.hyp <- c(5,2,1,0,35,13,15,8)
> data.frame(smoking,obesity,snoring,n.tot,n.hyp)
  smoking obesity snoring n.tot n.hyp
1      No      No      No    60     5
2     Yes      No      No    17     2
3      No     Yes      No     8     1
4     Yes     Yes      No     2     0
5      No      No     Yes   187    35
6     Yes      No     Yes    85    13
7      No     Yes     Yes    51    15
8     Yes     Yes     Yes    23     8
```

The `gl` function, to "generate levels" is briefly introduced in Section 6.3. The three first arguments to `gl` are, respectively, the number of levels, the repeat count of each level, and the total length of the vector. A fourth argument can be used to specify the level names of the resulting factor. The result is apparent from the printout of the generated variables. They were put together in a data frame to get a nicer layout.

R is able to fit logistic regression analyses for tabular data in two different ways. You have to specify the response as a matrix, where one column is the number of "diseased" and the other is the number of "healthy" (or "success"/"failure", depending on context).

```
> hyp.tbl <- cbind(n.hyp,n.tot-n.hyp)
> hyp.tbl
     n.hyp
[1,]     5    55
[2,]     2    15
[3,]     1     7
[4,]     0     2
[5,]    35   152
[6,]    13    72
[7,]    15    36
```

```
[8,]    8    15
```

The cbind function ("c" for "column") is used to bind variables together, columnwise, to form a matrix. Note that it would be a horrible mistake to use the total count for column 2 instead of the number of failures.

Then, you can specify the logistic regression model as

```
> glm(hyp.tbl~smoking+obesity+snoring,family=binomial("logit"))
```

Actually, "logit" is the default for binomial and the family argument is the second argument to glm, so it suffices to write

```
> glm(hyp.tbl~smoking+obesity+snoring,binomial)
```

The other way to specify a logistic regression model is to give the *proportion* of diseased in each cell:

```
> prop.hyp <- n.hyp/n.tot
> glm.hyp <- glm(prop.hyp~smoking+obesity+snoring,
+                binomial,weights=n.tot)
```

It is necessary to give weights because R cannot see how many observations a proportion is based on.

As output, you get in either case (except for minor details)

```
Call:  glm(formula = hyp.tbl ~ smoking + obesity + snoring, ...

Coefficients:
(Intercept)    smokingYes    obesityYes    snoringYes
   -2.37766      -0.06777      0.69531      0.87194

Degrees of Freedom: 7 Total (i.e. Null);   4 Residual
Null Deviance:       14.13
Residual Deviance: 1.618          AIC: 34.54
```

— which is in a minimal style, similar to that used for printing lm objects. Also in the result of glm is some nonvisible information, which may be extracted with particular functions. You can, for instance, save the result of a fit of a generalized linear model in a variable and obtain a table of regression coefficients and so forth, using summary:

```
> glm.hyp <- glm(hyp.tbl~smoking+obesity+snoring,binomial)
> summary(glm.hyp)

Call:
glm(formula = hyp.tbl ~ smoking + obesity + snoring, family ...

Deviance Residuals:
```

```
[1]   -0.04344    0.54145   -0.25476   -0.80051    0.19759   -0.46602
[7]   -0.21262    0.56231

Coefficients:
            Estimate Std. Error z value Pr(>|z|)
(Intercept) -2.37766    0.38012  -6.255 3.98e-10 ***
smokingYes  -0.06777    0.27811  -0.244   0.8075
obesityYes   0.69531    0.28508   2.439   0.0147 *
snoringYes   0.87194    0.39752   2.193   0.0283 *
---
Signif. codes:  0 '***' 0.001 '**' 0.01 '*' 0.05 '.' 0.1 ' ' 1

(Dispersion parameter for binomial family taken to be 1)

    Null deviance: 14.1259  on 7  degrees of freedom
Residual deviance:  1.6184  on 4  degrees of freedom
AIC: 34.537

Number of Fisher Scoring iterations: 3
```

In the following, we go through the components of summary output for
generalized linear models:

```
Call:
glm(formula = hyp.tbl ~ smoking + obesity + snoring, family = ...
```

As usual, we start off with a repeat of the model specification. Obviously,
more interesting is when the output is not viewed in connection with the
function call that generated it.

```
Deviance Residuals:
[1]   -0.04344    0.54145   -0.25476   -0.80051    0.19759   -0.46602
[7]   -0.21262    0.56231
```

This is the contribution of each cell of the table to the deviance of the
model (the deviance corresponds to the sum of squares in linear normal
models), with a sign according to whether the observation is larger or
smaller than expected. They can be used to pinpoint cells that are particularly poorly fitted, but you have to be wary of the interpretation in
sparse tables.

```
Coefficients:
            Estimate Std. Error z value Pr(>|z|)
(Intercept) -2.37766    0.38012  -6.255 3.98e-10 ***
smokingYes  -0.06777    0.27811  -0.244   0.8075
obesityYes   0.69531    0.28508   2.439   0.0147 *
snoringYes   0.87194    0.39752   2.193   0.0283 *
---
Signif. codes:  0 '***' 0.001 '**' 0.01 '*' 0.05 '.' 0.1 ' ' 1

(Dispersion parameter for binomial family taken to be 1)
```

This is the table of primary interest. Here, we get estimates of the regression coefficients, standard errors of same, and tests for whether each regression coefficient can be assumed to be zero. The layout is nearly identical to the corresponding part of the lm output.

The note about the dispersion parameter is related to the fact that the binomial variance depends entirely on the mean. There is no scale parameter like the variance in the normal distribution.

```
    Null deviance: 14.1259   on 7   degrees of freedom
Residual deviance:  1.6184   on 4   degrees of freedom
AIC: 34.537
```

"Residual deviance" corresponds to the residual sum of squares in ordinary regression analyses which is used to estimate the standard deviation about the regression line. In binomial models, however, the standard deviation of the observations is known, and you can therefore use the deviance in a test for model specification. The AIC (Akaike information criterion) is a measure of goodness of fit which takes the number of fitted parameters into account.

R is reluctant to associate a p-value with the deviance. Just as well, because no exact p-value can be found, only an approximation that is valid for large expected counts. In the present case, there are actually a couple of places where the expected cell count is rather small.

The asymptotic distribution of the residual deviance is a χ^2 distribution with the stated degrees of freedom, so even though the approximation may be poor, nothing in the data indicates that the model is wrong (the 5% significance limit is at 9.49 and the value found here is 1.62).

The null deviance is the deviance of a model that contains only the intercept, that is, describes a fixed probability (here: for hypertension) in all cells. What you would normally be interested in is the difference to the residual deviance, here $14.13 - 1.62 = 12.51$, which can be used for a joint test for whether any effects are present in the model. In the present case a p-value of approximately 6‰ is obtained.

```
Number of Fisher Scoring iterations: 3
```

This refers to the actual fitting procedure and is a purely technical item. There is no statistical information in it, but you should keep an eye on whether the number of iterations becomes too large, because that might be a sign that the model is too complex to fit based on the available data. Normally, glm halts the fitting procedure if the number of iterations exceeds 10, but it is possible to configure the limit.

The fitting procedure is *iterative*, in that there is no explicit formula that can be used to compute the estimates, only a set of equations that they should satisfy. However, there is an approximate solution of the equations if you supply an initial guess at the solution. This solution is then used as a starting point for an improved solution, and the procedure is repeated until the guesses are sufficiently stable.

A table of correlations between parameter estimates can be obtained via the optional argument corr=T to summary (this also works for linear models). It looks like this:

```
Correlation of Coefficients:
            (Intercept) smokingYes obesityYes
smokingYes     -0.1520
obesityYes     -0.1361 -9.499e-05
snoringYes     -0.8965 -6.707e-02  -0.07186
```

It is seen that the correlation between the estimates is fairly small, so that it may be expected that removing a variable from the model does not change the coefficients and *p*-values for other variables much. (The correlations between the regression coefficients and intercept are not very informative; they mostly relate to whether the variable in question has many or few observations in the "Yes" category.)

The z test in the table of regression coefficients immediately shows that the model can be simplified by removing smoking. The result then looks as follows (abbreviated):

```
> glm.hyp <- glm(hyp.tbl~obesity+snoring,binomial)
> summary(glm.hyp)
...
Coefficients:
            Estimate Std. Error z value Pr(>|z|)
(Intercept)  -2.3921     0.3757  -6.367 1.93e-10 ***
obesityYes    0.6954     0.2851   2.440   0.0147 *
snoringYes    0.8655     0.3966   2.182   0.0291 *
...
```

11.2.1 The analysis of deviance table

Deviance tables corresponds to ANOVA tables for multiple regression analyses and are generated like these with the anova function:

```
> glm.hyp <- glm(hyp.tbl~smoking+obesity+snoring,binomial)
> anova(glm.hyp, test="Chisq")
Analysis of Deviance Table

Model: binomial, link: logit
```

```
Response: hyp.tbl

Terms added sequentially (first to last)
```

	Df	Deviance	Resid. Df	Resid. Dev	P(>|Chi|)
NULL			7	14.1259	
smoking	1	0.0022	6	14.1237	0.9627
obesity	1	6.8274	5	7.2963	0.0090
snoring	1	5.6779	4	1.6184	0.0172

Notice that the Deviance column gives *differences* between models as variables are added to the model in turn. The deviances are approximately χ^2-distributed with the stated degrees of freedom. It is necessary to add the test="chisq" argument to get the approximate χ^2 tests.

Since the snoring variable on the last line is significant, it may not be removed from the model and we cannot use the table to justify model reductions. If, however, the terms are rearranged, so that smoking comes last, we get a deviance-based test for removal of that variable:

```
> glm.hyp <- glm(hyp.tbl~snoring+obesity+smoking,binomial)
> anova(glm.hyp, test="Chisq")
...
```

	Df	Deviance	Resid. Df	Resid. Dev	P(>|Chi|)
NULL			7	14.1259	
snoring	1	6.7887	6	7.3372	0.0092
obesity	1	5.6591	5	1.6781	0.0174
smoking	1	0.0597	4	1.6184	0.8069

From this, you can read that smoking is removable, whereas obesity is not, after removal of smoking.

For good measure, you should also set up the analysis with the two remaining explanatory variables interchanged, so that you get a test of whether snoring may be removed from a model that also contains obesity:

```
> glm.hyp <- glm(hyp.tbl~obesity+snoring,binomial)
> anova(glm.hyp, test="Chisq")
...
```

	Df	Deviance	Resid. Df	Resid. Dev	P(>|Chi|)
NULL			7	14.1259	
obesity	1	6.8260	6	7.2999	0.0090
snoring	1	5.6218	5	1.6781	0.0177

An alternative method is to use drop1 to try removing one term at a time:

```
> drop1(glm.hyp, test="Chisq")
Single term deletions
```

```
Model:
hyp.tbl ~ obesity + snoring
        Df Deviance    AIC    LRT Pr(Chi)
<none>        1.678 32.597
obesity  1    7.337 36.256  5.659 0.01737 *
snoring  1    7.300 36.219  5.622 0.01774 *
---
Signif. codes:  0 '***' 0.001 '**' 0.01 '*' 0.05 '.' 0.1 ' ' 1
```

Here, LRT is the likelihood ratio test, another name for the deviance change.

Actually, there is no more information in the deviance tables than in the z tests in the table of regression coefficients. From theoretical considerations, you might prefer the deviance test, but in practice the difference is small since $\chi^2 \approx z^2$ as long as you are looking at tests with a single degree of freedom. However, to test factors with more than two categories, it becomes unavoidable to use deviance tables, because the z tests relate only to some of the possible group comparisons.

11.2.2   Connection to test for trend

In Chapter 7 we consider tests for comparison of relative frequencies using prop.test and prop.trend.test, in particular the example of caesarean section versus shoe size. This example can also be analyzed as a logistic regression analysis on a "shoe score", which — for want of a better idea — may be chosen as the group number. This gives essentially the same analysis, in the sense that the same models are involved.

```
> data(caesarean)
> caesar.shoe
    <4  4 4.5  5 5.5  6+
Yes  5  7   6  7   8  10
No  17 28  36 41  46 140
> shoe.score <- 1:6
> shoe.score
[1] 1 2 3 4 5 6

> summary(glm(t(caesar.shoe)~shoe.score,binomial))
...
Coefficients:
            Estimate Std. Error z value Pr(>|z|)
(Intercept)  -0.8706     0.4051  -2.149  0.03161 *
shoe.score   -0.2597     0.0936  -2.775  0.00553 **
---
Signif. codes:  0 '***' 0.001 '**' 0.01 '*' 0.05 '.' 0.1 ' ' 1

(Dispersion parameter for binomial family taken to be 1)
```

```
    Null deviance: 9.3442  on 5  degrees of freedom
Residual deviance: 1.7845  on 4  degrees of freedom
AIC: 27.616
...
```

Notice that `caesar.shoe` had to be transposed with `t(...)`, so that the matrix was "stood on its end" in order to be used as the response variable by `glm`.

You can also write the results in a deviance table:

```
> anova(glm(t(caesar.shoe)~shoe.score,binomial))
...
              Df Deviance Resid. Df Resid. Dev
NULL                             5     9.3442
shoe.score     1   7.5597         4     1.7845
```

— from the last line of which you see that there is no significant deviation from linearity (1.78 on 4 degrees of freedom), whereas `shoe.score` has a significant contribution.

For comparison, the previous analyses using standard tests are repeated:

```
> caesar.shoe.yes <- caesar.shoe["Yes",]
> caesar.shoe.no <- caesar.shoe["No",]
> caesar.shoe.total <- caesar.shoe.yes+caesar.shoe.no
> prop.trend.test(caesar.shoe.yes,caesar.shoe.total)
        Chi-squared Test for Trend in Proportions
...
X-squared = 8.0237, df = 1, p-value = 0.004617
> prop.test(caesar.shoe.yes,caesar.shoe.total)

        6-sample test for equality of proportions without
        continuity correction
...
X-squared = 9.2874, df = 5, p-value = 0.09814
...
Warning message:
Chi-squared approximation may be incorrect in: prop.test(...
```

The 9.29 from `prop.test` corresponds to the 9.34 in residual deviance from a NULL model, whereas the 8.02 in the trend test corresponds to the 7.56 in the test of significance of `shoe.score`. Thus, the tests do not give exactly the same result, but generally *almost* the same. Theoretical considerations indicate that the specialized trend test is probably slightly better than the regression-based test. However, testing the linearity by subtracting the two χ^2 tests is definitely not as good as the real test for linearity.

11.3   Logistic regression using raw data

In this section we again use Anders Juul's data (see p. 73). For easy reference, here is how to read data and convert the variables that describe groupings into factors (this time slightly simplified):

```
> data(juul)
> juul$menarche <- factor(juul$menarche, labels=c("No","Yes"))
> juul$tanner <- factor(juul$tanner)
```

In the following we look at menarche as the response variable. This variable indicates for each girl whether or not she has had her first period. It is coded 1 for "no" and 2 for "yes". It is convenient to look at a subset of data consisting of 8–20-year-old girls. This can be extracted as follows:

```
> juul.girl <- subset(juul,age>8 & age<20 &
+                     complete.cases(menarche))
> attach(juul.girl)
```

For obvious reasons, no boys have a nonmissing menarche, so it is not necessary to select on gender explicitly.

Then you can analyze menarche as a function of age, like this:

```
> summary(glm(menarche~age,binomial))
Call:
glm(formula = menarche ~ age, family = binomial)

Deviance Residuals:
     Min        1Q    Median        3Q       Max
-2.32758  -0.18998   0.01253   0.12132   2.45922

Coefficients:
            Estimate Std. Error z value Pr(>|z|)
(Intercept) -20.0131     2.0216  -9.900   <2e-16 ***
age           1.5173     0.1539   9.862   <2e-16 ***
---
Signif. codes:  0 '***' 0.001 '**' 0.01 '*' 0.05 '.' 0.1 ' ' 1

(Dispersion parameter for binomial family taken to be 1)

    Null deviance: 719.39  on 518  degrees of freedom
Residual deviance: 200.66  on 517  degrees of freedom
AIC: 204.66

Number of Fisher Scoring iterations: 6
```

The response variable menarche is a *factor* with two levels, where the last level is considered the event. It also works to use a variable which has the values 0 and 1 (but *not*, for instance, 1 and 2!).

Notice that from this model you can estimate the median menarcheal age as the age where logit $p = 0$. A little thought (solve $-20.0131 + 1.5173 \times$ age $= 0$) reveals that it is $20.0131/1.5173 = 13.19$ years.

You should not pay too much attention to the deviance residuals in this case, since they automatically become large in every case where the fitted probability "goes against" the observations (which is bound to happen in some cases). The residual deviance is also difficult to interpret when there is only one observation per cell.

A hint of a more complicated analysis is obtained by including Tanner stage of puberty in the model. You should be warned that the exact interpretation of such an analysis is quite tricky and *qualitatively* different from the analysis of menarche as a function of age. It can be used for prediction purposes (although asking the girl whether she has had her first period would likely be much easier than determining her Tanner stage!), but the interpretation of the terms is not clearcut.

```
> summary(glm(menarche~age+tanner,binomial))
...
Coefficients:
            Estimate Std. Error z value Pr(>|z|)
(Intercept) -13.7727     2.7241  -5.056 4.28e-07 ***
age           0.8601     0.2283   3.768 0.000165 ***
tanner2      -0.5213     1.4717  -0.354 0.723179
tanner3       0.8263     1.2229   0.676 0.499222
tanner4       2.5644     1.2031   2.131 0.033052 *
tanner5       5.1889     1.3982   3.711 0.000206 ***
...
```

Notice that there is no joint test for the effect of `tanner`. There are a couple of significant z-values, so you would expect that the `tanner` variable has some effect (which, of course, you would probably expect even in the absence of data!). The formal test, however, must be obtained from the deviances:

```
> drop1(glm(menarche~age+tanner,binomial),test="Chisq")
...
       Df Deviance     AIC     LRT   Pr(Chi)
<none>     106.599 118.599
age     1  124.500 134.500  17.901 2.327e-05 ***
tanner  4  161.881 165.881  55.282 2.835e-11 ***
...
```

Clearly, both terms are highly significant.

11.4   Prediction

The predict function works for generalized linear models, too. Let's first consider the hypertension example, where data were given in tabular form:

```
> predict(glm.hyp)
[1] -2.3920762 -2.3920762 -1.6966574 -1.6966574 -1.5266180
[6] -1.5266180 -0.8311991 -0.8311991
```

Recall that smoking was eliminated from the model, which is why the expected values come in identical pairs.

These numbers are on the logit scale, which reveals the additive structure: Note that $2.392 - 1.697 = 1.527 - 0.831 = 0.695$ (except for roundoff error), which is exactly the regression coefficient to obesity. Likewise, the regression coefficient to snoring is obtained by looking at the differences $2.392 - 1.527 = 1.697 - 0.831 = 0.866$.

To get predicted values on the response scale (i.e., probabilities), use the type="response" argument to predict:

```
> predict(glm.hyp, type="response")
[1] 0.08377892 0.08377892 0.15490233 0.15490233 0.17848906
[6] 0.17848906 0.30339158 0.30339158
```

These may also be obtained using fitted, although you then cannot use the techniques for predicting on new data, etc.

In the analysis of menarche, the primary interest is probably in seeing a plot of the expected probabilities versus age (Figure 11.1). A crude plot could be obtained using something like

```
plot(age, fitted(glm(menarche~age,binomial)))
```

(it will look better if a different plotting symbol in a smaller size is used, using the pch and cex arguments) but here is a more ambitious plan:

```
> glm.menarche <- glm(menarche~age, binomial)
> Age <- seq(8,20,.1)
> newages <- data.frame(age=Age)
> predicted.probability <- predict(glm.menarche,
+                            newages,type="resp")
> plot(predicted.probability ~ Age, type="l")
```

Recall that seq generates equispaced vectors, here ages from 8 to 20 in steps of 0.1, so that connecting the points with lines will give a nearly smooth curve.

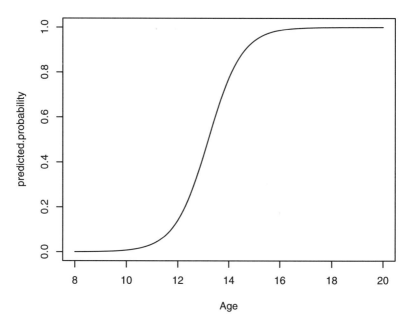

Figure 11.1. Fitted probability of menarche having occurred.

11.5 Model checking

For tabular data it is obvious to try to compare observed and fitted proportions. In the hypertension example you get

```
> fitted(glm.hyp)
[1] 0.08377892 0.08377892 0.15490233 0.15490233 0.17848906
[6] 0.17848906 0.30339158 0.30339158
> prop.hyp
[1] 0.08333333 0.11764706 0.12500000 0.00000000 0.18716578
[6] 0.15294118 0.29411765 0.34782609
```

The problem with this is that you get no feeling for how well the relative frequencies are determined. It can be better to look at observed and expected *counts* instead. The former can be computed as

```
> fitted(glm.hyp)*n.tot
[1]  5.0267353  1.4242417  1.2392186  0.3098047 33.3774535
[6] 15.1715698 15.4729705  6.9780063
```

— and to get a nice print for the comparison, you can use

```
> data.frame(fit=fitted(glm.hyp)*n.tot,n.hyp,n.tot)
         fit n.hyp n.tot
1  5.0267353     5    60
2  1.4242417     2    17
3  1.2392186     1     8
4  0.3098047     0     2
5 33.3774535    35   187
6 15.1715698    13    85
7 15.4729705    15    51
8  6.9780063     8    23
```

Notice that the discrepancy in cell 4 between 15% expected and 0% observed really is that there are 0 hypertensive out of 2 in a cell where the model yields an expectation of 0.3 hypertensives!

For complex models with continuous background variables, it becomes more difficult to perform an adequate model check. It is especially a hindrance that nothing really corresponds to a residual plot when the observations have only two different values.

Let's consider the example of the probability of menarche as a function of age. The problem here is whether the relation can really be assumed linear on the logit scale. For this case, you might try subdividing the x-axis in a number of intervals and see how the counts in each interval fit with the expected probabilities. This is presented graphically in Figure 11.2. Notice that the code *adds* points to Figure 11.1, which you are assumed not to have deleted at this point.

```
> age.group <- cut(age,c(8,10,12,13,14,15,16,18,20))
> tb <- table(age.group,menarche)
> tb
           menarche
age.group  No Yes
   (8,10]  100   0
  (10,12]   97   4
  (12,13]   32  21
  (13,14]   22  20
  (14,15]    5  36
  (15,16]    0  31
  (16,18]    0 105
  (18,20]    0  46
> rel.freq <- prop.table(tb,1)[,2]
> rel.freq
      (8,10]      (10,12]      (12,13]      (13,14]      (14,15]
0.00000000 0.03960396 0.39622642 0.47619048 0.87804878
      (15,16]      (16,18]      (18,20]
1.00000000 1.00000000 1.00000000
> points(rel.freq ~ c(9,11,12.5,13.5,14.5,15.5,17,19),pch=5)
```

The technique used above probably requires some explanation. First, cut is used to define the factor age.group, which describes a grouping

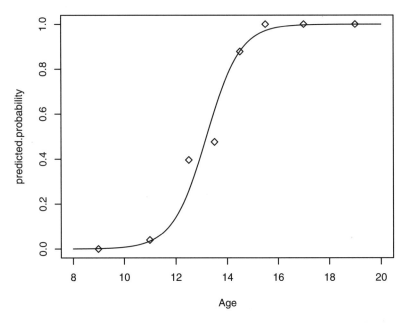

Figure 11.2. Fitted probability for menarche having occurred and observed proportion in age groups.

into age intervals. Then a crosstable `tb` is formed from `menarche` and `age.group`. Using `prop.table`, the numbers are expressed relative to the row total, and column 2 of the resulting table is extracted. This contains the relative proportion in each age group of girls with menarche having occurred. Finally, a plot of expected probabilities is made, overlaid by the observed proportions.

The plot looks reasonable on the whole, although the observed proportion among 12–13-year-olds appears a bit high and the proportion among 13–14-year-olds is a bit too low.

But how do you evaluate whether the deviation is larger than what can be expected from the statistical variation? One thing to try is to extend the model with a factor that describes a division into intervals. It is not practical to use the full division of `age.group`, because there are cells where either none or all of the girls have had their menarche.

We therefore try a division into four groups, with cutpoints at 12, 13, and 14 years, and add this factor to the model containing a linear age effect.

```
> age.gr <- cut(age,c(8,12,13,14,20))
```

```
> summary(glm(menarche~age+age.gr,binomial))
...
Coefficients:
                Estimate Std. Error z value Pr(>|z|)
(Intercept)     -21.5680     5.0462  -4.274 1.92e-05 ***
age               1.6250     0.4400   3.693 0.000222 ***
age.gr(12,13]     0.7296     0.7847   0.930 0.352450
age.gr(13,14]    -0.5218     1.1163  -0.467 0.640155
age.gr(14,20]     0.2751     1.6036   0.172 0.863783
...

> anova(glm(menarche~age+age.gr,binomial))
...
        Df Deviance Resid. Df Resid. Dev
NULL                    518      719.39
age      1   518.73     517      200.66
age.gr   3     8.06     514      192.61
> 1-pchisq(8.058,3)
[1] 0.04482811
```

That is, the addition of the grouping actually does give a significantly
better deviance. The effect is not highly significant, but since the deviation concerns the ages where "much happens", you should probably be
cautious about postulating a logit-linear age effect.

Another possibility is to try a polynomial regression model. Here you
need at least a third-degree polynomial to describe the apparent stagna-
tion of the curve around 13 years of age. We do not look at this in great
detail, just see part of output and the graphical presentation of the model
in Figure 11.3.

```
> anova(glm(menarche~age+I(age^2)+I(age^3)+age.gr,binomial))
...
          Df Deviance Resid. Df Resid. Dev
NULL                      518      719.39
age        1   518.73     517      200.66
I(age^2)   1     0.05     516      200.61
I(age^3)   1     8.82     515      191.80
age.gr     3     3.34     512      188.46
Warning messages:
1: Algorithm did not converge in: (if (is.empty.model(mt))...
2: fitted probabilities numerically 0 or 1 occurred in: (if (...
3: Algorithm did not converge in: method(...
4: fitted probabilities numerically 0 or 1 occurred in: method(...
> glm.menarche <- glm(menarche~age+I(age^2)+I(age^3), binomial)
Warning messages:
1: Algorithm did not converge in: (if (is.empty.model(...
2: fitted probabilities numerically 0 or 1 occurred in: (if (...
> predicted.probability <-
+     predict(glm.menarche, newages, type="resp")
> plot(predicted.probability ~ Age, type="l")
> points(rel.freq~c(9,11,12.5,13.5,14.5,15.5,17,19), pch=5)
```

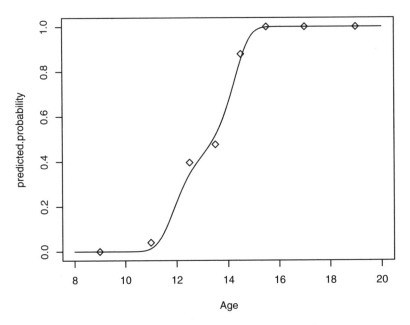

Figure 11.3. Logit-cubical fit of menarche data.

The warnings about fitted probabilities of 0 or 1 occur because the cubic term makes the logit tend much faster to $\pm\infty$ than the linear model did. The warning of nonconvergence is in this case simply due to the poor starting value ($p = 0.5$ for all observations), causing the algorithm to reach the maximum iteration count of 10. Increasing the maximum yields convergence in 11 steps, with values that do not differ materially from those above. This can be obtained using `glm(....,control=glm.control(maxit=20))`.

The thing to note in the deviance table is that the cubic term gives a substantial improvement of the deviance, but once that is included, the age grouping gives no additional improvement. The plot should speak for itself.

11.6   Exercises

11.1 In the `malaria` data set analyze the risk of malaria with age and log-transformed antibody level as explanatory variables.

11.2 Fit a logistic regression model to the graft.vs.host data set, predicting the gvhd response. Use different transformations of the index variable. Reduce the model using backward elimination.

11.3 In the analyses of the malaria and graft.vs.host data, try using the confint function from the MASS package to find improved confidence intervals for the regression coefficients.

11.4 Following up on Exercise 7.2 about "Rocky Mountain spotted fever", splitting the data by age groups gives the table below. Does this confirm the earlier analysis?

| | Western type | | Eastern type | |
Age group	Total	Fatal	Total	Fatal
Under 15	108	13	310	40
15–39	264	40	189	21
40 or above	375	157	162	61
	747	210	661	122

11.5 A *probit* regression is just like a logistic regression, but using a different link function. Try the analysis of the menarche variable in the juul data set with this link. Does the fit improve?

12

Survival analysis

The analysis of lifetimes is an important topic within biology and medicine in particular, but also in reliability analysis with engineering applications. Such data are often highly nonnormally distributed, so that the use of standard linear models is problematic.

Lifetime data are often *censored:* You do not know the exact lifetime, only that it is longer than a given value. For instance, in a cancer trial some people are lost to follow-up or simply live beyond the study period. It is an error to ignore the censoring in the statistical analysis, sometimes with extreme consequences. Consider, for instance, the case where a new treatment is introduced toward the end of the study period, so that nearly all the observed lifetimes will be cut short.

12.1   Essential concepts

Let X be the true lifetime and T a censoring time. What you observe is the minimum of X and T together with an indication of whether it is one or the other. T can be a random variable or a fixed time depending on context, but if it is random then it should generally be *noninformative* for the methods we describe here to be applicable. Sometimes "dead from other causes" is considered a censoring event for the mortality of a given disease, and in those cases it is particularly important to ensure that these other causes are unassociated with the disease state.

The *survival function* $S(t)$ measures the probability of being alive at a given time. It is really just 1 minus the cumulative distribution function for X, $1 - F(t)$.

The *hazard function* or *force of mortality* $h(t)$ measures the (infinitesimal) risk of dying within a short interval of time t, given that the subject is alive at time t. If the lifetime distribution has density f, then $h(t) = f(t)/S(t)$. This is often considered a more fundamental quantity than (say) the mean or median of the survival distribution and used as a basis for modelling.

12.2   Survival objects

We use the package called `survival` written by Terry Therneau and ported to R by Thomas Lumley. The package implements a large number of advanced techniques. For the present purposes, we use only a small subset of it.

To load `survival`, use

```
> library(survival)
```

The routines in `survival` work with objects of class `Surv`, which is a data structure that combines times and censoring information. Such objects are constructed using the `Surv` function, which takes two arguments: an observation time and an event indicator. The latter can be coded as a logical variable, a 0/1 variable, or a 1/2 variable. The latter coding is not recommended since `Surv` will assume 0/1 coding if all values are 1.

Actually, `Surv` can also be used with three arguments for dealing with data that have a start time as well as an end time ("staggered entry") and also interval censored data (where you know that an event happened between two dates, as happens, for instance, in repeated testing for a disease) can be handled.

We use the data set `melanom` collected by K.T. Drzewiecki and reproduced in Andersen et al. (1991). The data become accessible as follows

```
> data(melanom)
> attach(melanom)
> names(melanom)
[1] "no"      "status" "days"    "ulc"     "thick"  "sex"
```

The variable `status` is an indicator of the patient's status by the end of the study: 1 means "dead from malignant melanoma", 2 means "alive on January 1, 1978", and 3 means "dead from other causes". The variable `days` is the observation time in days, `ulc` indicates (1/2 for present/absent) whether the tumor was ulcerated, `thick` is the thickness in 1/100 mm, and `sex` contains the gender of the patient (1 for women and 2 for men).

We want to create a `Surv` object in which we consider the values 2 and 3 of the `status` variable as censorings. This is done as follows:

```
> Surv(days, status==1)
  [1]    10+    30+    35+    99+   185    204    210    232   232+   279
 [11]   295   355+   386    426   469    493+   529    621    629    659
 [21]   667   718    752    779   793    817    826+   833    858    869
...
[181] 3476+ 3523+ 3667+ 3695+ 3695+ 3776+ 3776+ 3830+ 3856+ 3872+
[191] 3909+ 3968+ 4001+ 4103+ 4119+ 4124+ 4207+ 4310+ 4390+ 4479+
[201] 4492+ 4668+ 4688+ 4926+ 5565+
```

Associated with the `Surv` objects is a print method that displays the objects in the above format, with a '+' marking censored observations. For example, `10+` means that the patient did not die from melanoma within 10 days and was then unavailable for further study (in fact, he died from other causes), whereas `185` means that the patient died from the disease a little over half a year after his operation.

Notice that the second argument to `Surv` is a logical vector; `status==1` is TRUE for those who died of malignant melanoma and FALSE otherwise.

12.3   Kaplan–Meier estimates

The Kaplan–Meier estimator allows the computation of an estimated survival function in the presence of right-censoring. It is also called the *product-limit estimator* because one way of describing the procedure is that it multiplies together conditional survival curves for intervals in which there are either no censored observations or no deaths. This becomes a step function where the estimated survival is reduced by a factor $(1 - 1/R_t)$ if there is a death at time t and a population of R_t is still alive and uncensored at that time.

Computing the Kaplan–Meier estimator for the survival function is done with a function called `survfit`. In its simplest form it takes just a single argument, namely a `Surv` object. It returns a `survfit` object. As described above, we consider "dead from other causes" a kind of censoring and do as follows:

```
> survfit(Surv(days,status==1))
Call: survfit(formula = Surv(days, status == 1))

      n   events    rmean se(rmean)    median  0.95LCL  0.95UCL
    205       57     4125       161       Inf      Inf      Inf
```

As can be seen, using `survfit` by itself is not very informative (just like the printed output of a "bare" `lm` is not). You get a couple of summary statistics and an estimate of the median survival, and in this case the latter is not even interesting because the estimate is infinite. The survival curve does not cross the 50% mark before all patients were censored. The finite estimate for the mean arises from setting the survival function to zero beyond the last observation time (thus the label `rmean`).

To see the actual Kaplan–Meier estimate, use `summary` on the `survfit` object. We first save the `survfit` object into a variable, here named `surv.all` because it contains the raw survival function for all patients without regard to patient characteristics.

```
> surv.all <- survfit(Surv(days,status==1))
> summary(surv.all)
Call: survfit(formula = Surv(days, status == 1))

 time n.risk n.event survival std.err lower 95% CI upper 95% CI
  185    201       1    0.995 0.00496        0.985        1.000
  204    200       1    0.990 0.00700        0.976        1.000
  210    199       1    0.985 0.00855        0.968        1.000
  232    198       1    0.980 0.00985        0.961        1.000
  279    196       1    0.975 0.01100        0.954        0.997
  295    195       1    0.970 0.01202        0.947        0.994
 ...
 2565     63       1    0.689 0.03729        0.620        0.766
 2782     57       1    0.677 0.03854        0.605        0.757
 3042     52       1    0.664 0.03994        0.590        0.747
 3338     35       1    0.645 0.04307        0.566        0.735
```

This contains the value of the survival function at the event times. The censoring times are not displayed but are contained in the `survfit` object and can be obtained by passing `censored=T` to `summary` (see the help page for `summary.survfit` for such details).

The Kaplan–Meier estimate is the step function whose jump points are given in `time` and whose values right after a jump are given in `survival`. Additionally, both an estimate of the standard error of the curve and a (pointwise) confidence interval for the true curve are given.

Normally, you would be more interested in showing the Kaplan–Meier estimate graphically than numerically. To do this (Figure 12.1), you simply write

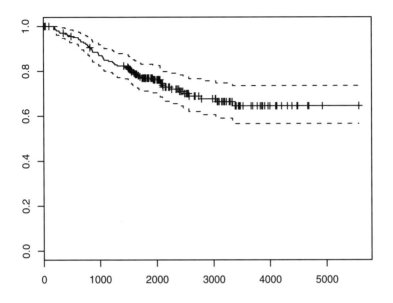

Figure 12.1. Kaplan–Meier plot for melanoma data (all observations).

```
> plot(surv.all)
```

The markings on the curve indicate censoring times, and the bands give approximate confidence intervals. If you look closely, you will see that the bands are not symmetrical around the estimate. They are constructed as a symmetric interval on the log scale and transformed back to the original scale.

It is often useful to plot two or more survival functions on the same plot so that they can be directly compared (Figure 12.2). To obtain survival functions split by gender, do the following:

```
> surv.bysex <- survfit(Surv(days,status==1)~sex)
> plot(surv.bysex)
```

That is, you use a model formula as in lm and glm, specifying that the survival object generated from day and status should be described by sex. Notice that there are no confidence intervals on the curves. These are turned off when there are two or more curves, because the display easily becomes confusing. They can be turned on again by passing conf.int=T to plot, in which case it can be recommended to use separate colors for the curves, as in

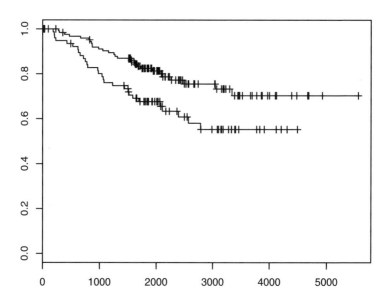

Figure 12.2. Kaplan–Meier plots for melanoma data, grouped by gender.

```
> plot(surv.bysex, conf.int=T, col=c("black","gray"))
```

Similarly, you can avoid plotting the confidence bands in the single sample case by setting `conf.int=F`. If you want the bands but at a 99% confidence level, you should pass `conf.int=0.99` to `survfit`. Notice that the level of confidence is an argument to the fitting function (which needs it to compute the confidence limits), whereas the decision to plot the bands is controlled by a similarly named argument to plot.

12.4   The log-rank test

The log-rank test is used to test whether two or more survival curves are identical. It is based on looking at the population at each death time and computing the expected number of deaths in proportion to the number of individuals at risk in each group. This is then summed over all death times and compared to the observed number of deaths by a procedure similar (but not identical) to the χ^2 test. Notice that the interpretation of "expected" and "observed" is slightly peculiar: If the difference in mortality is sufficiently large, then you can easily "expect" the same individuals to die several times over the course of the trial. If the population is observed to extinction with no censoring, then the observed number of deaths will equal the group size by definition and the expected values will contain all the random variation.

The log-rank test is formally nonparametric, since the distribution of the test statistic depends only on the assumption that the groups have the same survival function. However, it can also be viewed as a model-based test under the assumption of *proportional hazards* (see Section 12.1). You can set up a semiparametric model in which the hazard itself is unspecified but it is assumed that the hazards are proportional between groups. Testing that the proportionality factors are all unity then leads to a log-rank test. The log-rank test will work best against this class of alternatives.

Computing the log-rank test is done by the function survdiff. This actually implements a whole family of tests specified by a parameter *rho*, allowing various nonproportional hazards alternatives to the null hypothesis, but the default value of *rho* = 0 gives the log-rank test.

```
> survdiff(Surv(days,status==1)~sex)
Call:
survdiff(formula = Surv(days, status == 1) ~ sex)

        N Observed Expected (O-E)^2/E (O-E)^2/V
sex=1 126       28     37.1      2.25      6.47
sex=2  79       29     19.9      4.21      6.47

 Chisq= 6.5  on 1 degrees of freedom, p= 0.011
```

The specification is using a model formula as for linear and generalized linear models. However, the test can deal only with grouped data, so if you specify multiple variables on the right-hand side it will work on the grouping of data generated by all combinations of predictor variables. It also makes no distinction between factors and numerical codes. The same is true of survfit.

It is also possible to specify stratified analyses, in which the observed and expected value calculations are carried out separately within a stratification of the data set. For instance, you can compute the log-rank test for a gender effect stratified by ulceration as follows:

```
> survdiff(Surv(days,status==1)~sex+strata(ulc))
Call:
survdiff(formula = Surv(days, status == 1) ~ sex + strata(ulc))

        N Observed Expected (O-E)^2/E (O-E)^2/V
sex=1 126       28     34.7      1.28      3.31
sex=2  79       29     22.3      1.99      3.31
```

```
Chisq= 3.3   on 1 degrees of freedom, p= 0.0687
```

Notice that this makes the effect of sex appear less significant. A possible explanation might be that males seek treatment when the disease is in a more advanced state than women do, so that the gender difference is reduced when adjusted for a measure of disease progression.

12.5   The Cox proportional hazards model

The proportional hazards model allows the analysis of survival data by regression models similar to those of lm and glm. The scale on which linearity is assumed is the log-hazard scale. Models can be fitted via the maximization of *Cox's likelihood*, which is not a true likelihood, but it can be shown that it may be used as one. It is calculated in a matter similar to the log-rank test, as the product of conditional likelihoods of the observed death at each death time.

As a first example, consider a model with the single regressor sex:

```
> summary(coxph(Surv(days,status==1)~sex))
Call:
coxph(formula = Surv(days, status == 1) ~ sex)

  n= 205

      coef exp(coef) se(coef)    z    p
sex 0.662      1.94    0.265 2.50 0.013

      exp(coef) exp(-coef) lower .95 upper .95
sex        1.94      0.516      1.15      3.26

Rsquare= 0.03    (max possible= 0.937 )
Likelihood ratio test= 6.15   on 1 df,    p=0.0131
Wald test            = 6.24   on 1 df,    p=0.0125
Score (logrank) test = 6.47   on 1 df,    p=0.0110
```

The coef is the estimated logarithm of the hazard ratio between the two groups, which for convenience is also given as the actual hazard ratio exp(coef). The line following that also gives the inverted ratio (swapping the groups) and confidence intervals·for the hazard ratio. Finally, three overall tests for significant effects in the model are given. These are all equivalent in large samples but may differ somewhat in small-sample cases. Notice that the Wald test is identical to the z test based on the estimated coefficient divided by its standard error, whereas the score test is equivalent to the log-rank test (as long as the model involves only a simple grouping).

A more elaborate example, involving a continuous covariate and a stratification variable, is

```
> summary(coxph(Surv(days,status==1)~sex+log(thick)+strata(ulc)))
Call:
coxph(formula = Surv(days, status == 1) ~ sex + log(thick) +
    strata(ulc))

  n= 205

              coef exp(coef) se(coef)    z      p
sex           0.36    1.43     0.270 1.33 0.1800
log(thick) 0.56       1.75     0.178 3.14 0.0017

           exp(coef) exp(-coef) lower .95 upper .95
sex             1.43     0.698     0.844      2.43
log(thick)      1.75     0.571     1.234      2.48

Rsquare= 0.063    (max possible= 0.9 )
Likelihood ratio test= 13.3  on 2 df,    p=0.00130
Wald test             = 12.9  on 2 df,    p=0.0016
Score (logrank) test = 13.0  on 2 df,    p=0.00152
```

It is seen that the significance of the sex variable has been further reduced.

The Cox model assumes an underlying baseline hazard function, with a corresponding survival curve. In a stratified analysis there will be one such curve for each stratum. They can be extracted by using survfit on the output of coxph and of course be plotted using the plot method for survfit objects (Figure 12.3):

```
> plot(survfit(coxph(Surv(days,status==1)~
+                log(thick)+sex+strata(ulc)))))
```

Beware that the default for survfit is to generate curves for a pseudo-individual for which the covariates are at their mean value. In the present case that would correspond to a tumor thickness of 1.86 mm and a gender of 1.39 (!). Notice that we have been sloppy in not defining sex as a factor variable, but that would not actually give a different result (coxph subtracts the mean of the regressors before fitting, so a 1/2 coding is the same as 0/1, which is what a factor with treatment contrasts gives you). However, you can use the newdata argument of survfit to specify a data frame for which you want to calculate survival curves.

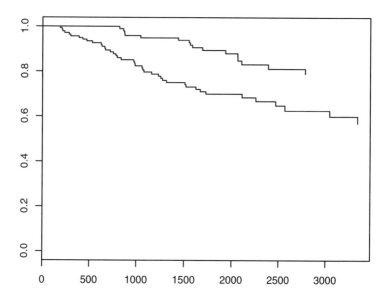

Figure 12.3. Baseline survival curves (ulcerated and nonulcerated tumors) in stratified Cox regression.

12.6   Exercises

12.1 In the graft.vs.host data set estimate the survival function for patients with or without GVHD. Test the hypothesis that the survival is the same in both groups. Extend the analysis by including the other explanatory variables.

12.2 With the Cox model in the last section of the text, generate a plot with estimated survival curves for men with nonulcerated tumors of thicknesses 0.1, 0.2, and 0.5 mm (three curves in one plot). Hint: survfit objects can be indexed with [] to extract individual strata.

A

Obtaining and installing R

The way to obtain R is to download it from one of the CRAN (Comprehensive R Archive Network) sites. The main site is

```
http://cran.r-project.org/
```

It has a number of mirror sites worldwide, which may be closer to you and give faster download times.

Installation details tend to vary over time, so you should read the accompanying documents and any other information offered on CRAN.

Binary distributions

As of this writing, the version for recent variants of Microsoft Windows comes as a single SetupR.exe file, on which you simply double-click with the mouse and then follow the on-screen instructions. When the process is completed, you will have an entry under Programs on the Start menu for invoking R, as well as a desktop icon.

For Linux distributions that use the RPM package format (RedHat, Mandrake, LinuxPPC, and SuSE) and also for Alpha Unix (OSF/Tru64), .rpm files of R and the recommended add-on packages can be installed using the rpm command. Packages for the Debian APT package manager are also available.

For the Macintosh platforms there are two different binary distributions: the "Carbon" R and the "Darwin" R. The Carbon version is intended to run natively on MasOS System from 8.6 to OS X, and the Darwin one as a usual Unix command under OS X. The Darwin R also requires an X windows manager like XDarwin to use the X11 graphic device.

Carbon R comes in single .sit archive file that you simply decompress by dragging the file onto Stuffit Expander and move the resulting folder rmxyz into your favourite applications folder. The Darwin version is a .tgz archive, which can be installed, after decompression, with some (fairly trivial) manual adjustments.

Darwin R can also be installed using the "fink". Fink installs all dynamic libraries that might be needed, and it can update R to newer versions when available.

Installation from source

Installation from source code is possible on all supported platforms, although nontrivial on Macintosh and Windows, mainly because the build environment is not part of the system. On Unix-like systems (Macintosh OS X included), the process can be as simple as unpacking the sources and writing

```
./configure
make
make install
```

and then you would unpack the recommended package bundle, change to its directory and enter

```
R CMD INSTALL *.tar.gz
```

The above works on widely used platforms, provided that the relevant compilers and support libraries are installed. If your system is more esoteric or you want to use special compilers or libraries, then you may need to dig deeper.

For Windows and Carbon Macintosh, the directories src/gnuwin32 and src/macintosh have an INSTALL file with detailed information about the procedure to follow.

Package installation

To install the ISwR package with data sets used in this book under Unix/Linux or Windows, you can connect to the Internet, start R, and enter

```
install.packages("ISwR", .libPaths()[1])
```

The Windows version provides a convenient menu interface for the operation.

If your R machine is not connected to the Internet, you can also download the package as a file and install that. For Windows and the Carbon version for Macintosh, you need to get the binary package (.zip or .sit extension). For Windows, installation from a local .zip file is possible via a menu entry. For Macintosh users, the procedure is described in the Macintosh FAQ. For Unix and Linux, you can issue the following at the shell prompt (the -l option allows you to give a private library):

```
R CMD INSTALL ISwR
```

On Unix and Linux systems you will need superuser permissions to install. Otherwise you can set up a private library directory and install into that. Use the R_LIBS environment variable to use your private library subsequently. A similar issue arises if R is installed on a read-only file system in a Windows environment. Further details can be found in the help page for library.

More information

Information and further Internet resources for R can be obtained from CRAN and the R homepage at

```
www.r-project.org
```

Notice in particular the mailing lists, the user-contributed documents, and the FAQs.

B

Data sets in the ISwR package[1]

ashina	*Ashina's crossover trial*

Description

The ashina data frame has 16 rows and 3 columns. It contains data from a crossover trial for the effect of an NO synthase inhibitor on headaches. Visual analog scale recordings of pain level were made at baseline and at five time points after infusion of the drug or placebo. A score was calculated as the sum of the differences from baseline. All patients received both treatment and placebo in randomized order.

Format

This data frame contains the following columns:

vas.active a numeric vector. Summary score when given active substance.

vas.plac a numeric vector. Summary score when given placebo treatment.

grp a numeric vector code. 1: placebo first, 2: active first.

---

[1]Reproduced with permission from the documentation files in the ISwR package

Source

Original data

References

M.Ashina et al. (1999), *Lancet* 353, pp. 287–289

Examples

```
data(ashina)
plot(vas.active~vas.plac,pch=grp,data=ashina)
abline(0,1)
```

---

bp.obese   *Obesity and blood pressure*

Description

The bp.obese data frame has 102 rows and 3 columns. It contains
data from a random sample of Mexican-American adults in a small
California town.

Format

This data frame contains the following columns:

sex a numeric vector code. 0: male, 1: female.
obese a numeric vector. Ratio of actual weight to ideal weight from
New York Metropolitan Life Tables.
bp a numeric vector. Systolic blood pressure (mm Hg).

Source

B.W. Brown & M. Hollander (1977), Statistics. A Biomedical Introduction, Wiley.

Examples

```
data(bp.obese)
plot(bp~obese,pch = ifelse(sex==1, "F", "M"), data = bp.obese)
```

caesarean	*Caesarean section and maternal shoe size*

Description

The table `caesar.shoe` contains the relation between caesarean section and maternal shoe sizes (UK sizes!)

Format

A matrix with two rows and six columns

Source

D.G. Altman (1991), *Practical Statistics for Medical Research*, Table 10.1, Chapman & Hall.

Examples

```
data(caesarean)
require(ctest)
prop.trend.test(caesar.shoe["Yes",],margin.table(caesar.shoe,2))
```

coking	*Coking data*

Description

The `coking` data frame has 18 rows and 3 columns. It contains the time to coking in an experiment where the oven width and temperature were varied.

Format

This data frame contains the following columns:

width a factor with levels 4, 8, and 12, giving the oven width in inches.

temp a factor with levels 1600 and 1900, giving the temperature in Fahrenheit.

time a numeric vector, time to coking.

Source

R.A. Johnson (1994), *Miller and Freund's Probability and Statistics for Engineers*, 5th ed., Prentice-Hall.

Examples

```
data(coking)
attach(coking)
matplot(tapply(time,list(width,temp),mean))
detach(coking)
```

---

cystfibr   *Cystic fibrosis lung function data*

Description

The cystfibr data frame has 25 rows and 10 columns. It contains lung function data for cystic fibrosis patients (7–23 years old)

Format

This data frame contains the following columns:

age a numeric vector. Age in years.
sex a numeric vector code. 0: male, 1:female.
height a numeric vector. Height (cm).
weight a numeric vector. Weight (kg).
bmp a numeric vector. Body mass (% of normal).
fev1 a numeric vector. Forced expiratory volume.
rv a numeric vector. Residual volume.
frc a numeric vector. Functional residual capacity.
tlc a numeric vector. Total lung capacity.
pemax a numeric vector. Maximum expiratory pressure.

Source

D.G. Altman (1991), *Practical Statistics for Medical Research*, Table 12.11, Chapman & Hall.

References

O'Neill et al. (1983) (full reference in Altman).

---

energy   *Energy expenditure*

Description

The energy data frame has 22 rows and 2 columns. It contains data
on the energy expenditure in groups of lean and obese women.

Format

This data frame contains the following columns:

expend a numeric vector. 24 hour energy expenditure (MJ).
stature a factor with levels lean and obese.

Source

D.G. Altman (1991), *Practical Statistics for Medical Research*, Table 9.4,
Chapman & Hall.

Examples

```
data(energy)
plot(expend~stature,data=energy)
```

---

fake.trypsin   *Trypsin by age groups*

Description

The trypsin data frame has 271 rows and 3 columns. Serum levels
of immunoreactive trypsin in healthy volunteers (faked!).

Format

This data frame contains the following columns:

trypsin a numeric vector. Serum-trypsin in ng/ml.
grp a numeric vector. Age coding. See below.
grpf a factor with levels 1: age 10–19, 2: age 20–29, 3: age 30–39, 4:
age 40–49, 5: age 50–59, and 6: age 60–69.

Details

Data have been simulated to match given group means and SD.

Source

D.G. Altman (1991), *Practical Statistics for Medical Research*, Table 9.12, Chapman & Hall.

Examples

```
data(fake.trypsin)
plot(trypsin~grp, data=fake.trypsin)
```

---

graft.vs.host   *Graft versus host disease*

Description

The gvhd data frame has 37 rows and 7 columns. It contains data from patients receiving a nondepleted allogenic bone marrow transplant, with the purpose of finding variables associated with the development of acute graft-versus-host disease.

Format

This data frame contains the following columns:

pnr a numeric vector. Patient number.

rcpage a numeric vector. Age of recipient (years).

donage a numeric vector. Age of donor (years).

type a numeric vector, type of leukaemia coded 1: AML, 2: ALL, 3: CML for acute myeloid, acute lymphatic, and chronic myeloid leukaemia.

preg a numeric vector code, indicating whether donor has been pregnant. 0: no, 1: yes.

index a numeric vector giving an index of mixed epidermal cell-lymphocyte reactions.

gvhd a numeric vector code, graft versus host disease. 0: no, 1: yes.

time a numeric vector. Follow-up time

dead a numeric vector code 0: no (censored), 1: yes

Source

D.G. Altman (1991), *Practical Statistics for Medical Research*, Exercise 12.3, Chapman & Hall.

Examples

```
data(graft.vs.host)
plot(jitter(gvhd,0.2)~index,data=graft.vs.host)
```

---

heart.rate   *Heart rates after enalaprilat*

Description

The heart.rate data frame has 36 rows and 3 columns. It contains data for nine patients with congestive heart failure before and shortly after administration of enalaprilat, in a balanced two-way layout.

Format

This data frame contains the following columns:

hr a numeric vector. Heart rate in beats per minute.
subj a factor with levels 1 to 9.
time a factor with levels 0 (before), 30, 60, and 120 (minutes after administration).

Source

D.G. Altman (1991), *Practical Statistics for Medical Research*, Table 12.2, Chapman & Hall.

Examples

```
data(heart.rate)
evalq(interaction.plot(time,subj,hr), heart.rate)
```

---

hellung   *Growth of Tetrahymena cells*

Description

The hellung data frame has 51 rows and 3 columns. diameter and concentration of *Tetrahymena* cells, with and without glucose added to growth medium.

Format

This data frame contains the following columns:

glucose a numeric vector code 1: yes, 2: no.
conc a numeric vector. Cell condentration (counts/ml)
diameter a numeric vector. Cell diameter (μm)

Source

D. Kronborg and L.T. Skovgaard (1990), *Regressionsanalyse*, Table 1.1, FADLs Forlag (in Danish).

Examples

```
data(hellung)
plot(diameter~conc,pch=glucose,log="xy",data=hellung)
```

IgM   *Immunoglobulin G*

Description

Serum IgM in 298 children aged 6 months to 6 years.

Format

A single numeric vector (g/l).

Source

D.G. Altman (1991), *Practical Statistics for Medical Research*, Table 3.2, Chapman & Hall.

Examples

```
data(IgM)
stripchart(IgM,method="stack")
```

intake	*Energy intake*

Description

The intake data frame has 11 rows and 2 columns. It contains paired values of energy intake for 11 women.

Format

This data frame contains the following columns:

pre a numeric vector. Premenstrual intake (kJ).
post a numeric vector. Postmenstrual intake (kJ).

Source

D.G. Altman (1991), *Practical Statistics for Medical Research*, Table 9.3, Chapman & Hall.

Examples

```
data(intake)
plot(intake$pre, intake$post)
```

juul	*Juul's IGF data, extended version*

Description

The juul data frame has 1339 rows and 6 columns. It contains a reference sample of the distribution of insulin-like growth factor (IGF-I), one observation per subject in various ages with the bulk of the data collected in connection with school physical examinations.

Format

This data frame contains the following columns:

age a numeric vector (years).
menarche a numeric vector. Has menarche occurred (code 1: no, 2: yes)?
sex a numeric vector (1: boy, 2: girl).
igf1 a numeric vector. Insulin-like growth factor ($\mu g/l$).

tanner a numeric vector. Codes 1–5: Stages of puberty a.m. Tanner.
testvol a numeric vector. Testicular volume (ml).

Source

Original data.

Examples

```
data(juul)
plot(igf1~age, data=juul)
```

---

juul2   *Juul's IGF data, extended version*

Description

The juul2 data frame has 1339 rows and 8 columns. Extended version of juul.

Format

This data frame contains the following columns:

age a numeric vector (years).
height a numeric vector (cm).
menarche a numeric vector. Has menarche occurred (code 1: no, 2: yes)?
sex a numeric vector (1: boy, 2: girl).
igf1 a numeric vector. Insulin-like growth factor (μg/l).
tanner a numeric vector. Codes 1–5: Stages of puberty a.m. Tanner.
testvol a numeric vector. Testicular volume (ml).
weight a numeric vector. Weight (kg).

Source

Original data.

Examples

```
data(juul2)
plot(igf1~age, data=juul2)
```

kfm	*Breast-feeding data*

Description

The kfm data frame has 50 rows and 7 columns. It was collected by Kim Fleischer Michaelsen and contains data for 50 infants of age approximately 2 months. They were weighed immediately before and after each breast feeding and the measured intake of breast milk was registered along with various other data.

Format

This data frame contains the following columns:

no a numeric vector. Identification number.
dl.milk a numeric vector. Breast-milk intake (dl/24h).
sex a factor with levels boy and girl
weight a numeric vector. Weight of child (kg).
ml.suppl a numeric vector. Supplementary milk substitute (ml/24h).
mat.weight a numeric vector. Weight of mother (kg).
mat.height a numeric vector. Height of mother (cm).

Note

The amount of supplementary milk substitute refers to a period before the data collection.

Source

Original data.

Examples

```
data(kfm)
plot(dl.milk~mat.height,pch=c(1,2)[sex],data=kfm)
```

lung	*Methods for determining lung volume*

Description

The lung data frame has 18 rows and 3 columns. It contains data on three different methods of determining human lung volume.

Usage

```
data(lung)
```

Format

This data frame contains the following columns:

volume a numeric vector. Measured lung volume.
method a factor with levels A, B, and C.
subject a factor with levels 1–6.

Source

Exercises in Applied statistics (1977), Exercise 4.15, Dept. of Theoretical Statistics, Aarhus University.

Examples

```
data(lung)
```

malaria	*Malaria antibody data*

Description

The malaria data frame has 100 rows and 4 columns.

Usage

```
data(malaria)
```

Format

This data frame contains the following columns:

subject subject code.
age age in years.
ab antibody level.
mal a numeric vector code: Malaria, 0/1 is no/yes, respectively.

Details

A random sample of 100 children from a village in Ghana, aged 3–15 years. The children were followed for a period of 8 months. At the beginning of the study, values of a particular antibody were assessed. Based on observations during the study period, the children were categorized into two groups: individuals with and without symptoms of malaria.

Source

Unpublished data.

Examples

```
data(malaria)
summary(malaria)
```

| melanom | *Survival after malignant melanoma* |

Description

The `melanom` data frame has 205 rows and 7 columns. It contains data relating to survival of patients after operation for malignant melanoma collected at Odense University Hospital by K.T. Drzewiecki.

Format

This data frame contains the following columns:

no a numeric vector. Patient code.
status a numeric vector code. Survival status. 1: dead from melanoma, 2: alive, 3: dead from other cause.
days a numeric vector. Observation time.
ulc a numeric vector code. Ulceration, 1: present, 2: absent.

thick a numeric vector. Tumor thickness (1/100 mm).
sex a numeric vector code. 1: female, 2: male.

Source

P.K. Andersen, Ø. Borgan, R.D. Gill, N. Keiding (1991), *Statistical Models Based on Counting Processes*, Appendix 1, Springer-Verlag.

Examples

```
data(melanom)
require(survival)
plot(survfit(Surv(days,status==1),data=melanom))
```

---

react   *Tuberculin reactions*

Description

The numeric vector react contains differences between two nurses' determination of 334 tuberculin reaction sizes.

Format

A single vector. Reaction sizes in mm.

Examples

```
data(react)
hist(react) # not good because of discretization effects...
plot(density(react))
```

---

red.cell.folate   *Red cell folate data*

Description

The folate data frame has 22 rows and 2 columns. It contains data on red cell folate levels in patients receiving three different methods of ventilation during anesthesia.

Format

This data frame contains the following columns:

folate a numeric vector. Folate concentration (μg/l).

ventilation a factor with levels N2O+O2, 24h: 50% nitrous oxide and 50% oxygen, continuously for 24 hours; N2O+O2, op: 50% nitrous oxide and 50% oxygen, only during operation; O2, 24h: no nitrous oxide, but 35–50% oxygen for 24 hours.

Source

D.G. Altman (1991), *Practical Statistics for Medical Research*, Table 9.10, Chapman & Hall.

Examples

```
data(red.cell.folate)
plot(folate~ventilation,data=red.cell.folate)
```

rmr	*Resting metabolic rate*

Description

The rmr data frame has 44 rows and 2 columns. It contains resting metabolic rate and body weight for 44 women.

Format

This data frame contains the following columns:

body.weight a numeric vector. Body weight (kg).

metabolic.rate a numeric vector. Metabolic rate (kcal/24 hr).

Source

D.G. Altman (1991), *Practical Statistics for Medical Research*, Exercise 11.2, Chapman & Hall.

Examples

```
data(rmr)
plot(metabolic.rate~body.weight,data=rmr)
```

---

secher   *Birth weight and ultrasonography*

Description

The secher data frame has 107 rows and 4 columns. It contains ultra-sonographic measurements of fetuses immediately before birth and subsequent birth weight.

Format

This data frame contains the following columns:

bwt a numeric vector. Birth weight (g).
bpd a numeric vector. Biparietal diameter (mm).
ad a numeric vector. Abdominal diameter (mm).
no a numeric vector. Observation number.

Source

D. Kronborg and L.T. Skovgaard (1990), *Regressionsanalyse*, Table 3.1, FADLs Forlag (in Danish).
Secher et al. (1987), Eur.j.obs.gyn.repr.biol., 24, 1–11.

Examples

```
data(secher)
plot(bwt~ad, data=secher, log="xy")
```

---

secretin   *Secretin-induced blood glucose changes*

Description

The secretin data frame has 50 rows and 6 columns. It contains data from a glucose response experiment.

Usage

```
data(secretin)
```

Format

This data frame contains the following columns:

gluc a numeric vector. Blood glucose level.
person a factor with levels A–E.
time a factor with levels 20, 30, 60, 90 (minutes since injection), and pre (before injection).
repl a factor with levels a: 1st sample and b: 2nd sample.
time20plus a factor with levels 20+: 20 minutes or longer since injection and pre: before injection.
time.comb a factor with levels 20: 20 minutes since injection, 30+: 30 minutes or longer since injection, and pre: before injection.

Details

Secretin is a hormone of the duodenal mucous membrane. An extract was administered to five patients with arterial hypertension. Primary registrations (double determination) of blood glucose were on graph paper, later quantified with the smallest of the two measurements recorded first.

Source

Exercises in Applied statistics (1977), Exercise 5.8, Dept. of Theoretical Statistics, Aarhus University.

Examples

```
data(secretin)
```

---

tb.dilute   *Tuberculin dilution assay*

Description

The `tb.dilute` data frame has 18 rows and 3 columns. It contains data from a drug test involving dilutions of tuberculin.

Usage

```
data(tb.dilute)
```

Format

This data frame contains the following columns:

reaction a numeric vector. Reaction sizes (average of diameters) for tuberculin skin pricks.
animal a factor with levels 1–6.
logdose a factor with levels 0.5, 0, and -0.5.

Details

The actual dilutions were 1:100, 1:100$\sqrt{10}$, 1:1000. Setting the middle one to 1 and using base-10 logarithms gives the logdose values.

Source

Exercises in Applied statistics (1977), part of Exercise 4.15, Dept. of Theoretical Statistics, Aarhus University.

Examples

```
data(tb.dilute)
```

---

thuesen   *Ventricular shortening velocity*

Description

The thuesen data frame has 24 rows and 2 columns. It contains ventricular shortening velocity and blood glucose for type 1 diabetic patients.

Format

This data frame contains the following columns:

blood.glucose a numeric vector. Fasting blood glucose (mmol/l).
short.velocity a numeric vector. Mean circumferential shortening velocity (%/s).

Source

D.G. Altman (1991), *Practical Statistics for Medical Research*, Table 11.6, Chapman & Hall.

Examples

```
data(thuesen)
plot(short.velocity~blood.glucose, data=thuesen)
```

---

tlc	*Total lung capacity*

Description

The tlc data frame has 32 rows and 4 columns. It contains data on pretransplant total lung capacity (TLC) for recipients of heart-lung transplants, by whole-body plethysmography.

Format

This data frame contains the following columns:

age a numeric vector. Age of recipient (years).
sex a numeric vector code. Female: 1, male: 2.
height a numeric vector. Height of recipient (cm).
tlc a numeric vector. Total lung capacity (l).

Source

D.G. Altman (1991), *Practical Statistics for Medical Research*, Exercise 12.5, 10.1, Chapman & Hall.

Examples

```
data(tlc)
plot(tlc~height,data=tlc)
```

---

vitcap	*Vital capacity*

Description

The vitcap data frame has 24 rows and 3 columns. It contains data on vital capacity for workers in the cadmium industry. It is a subset of the vitcap2 data set.

Format

This data frame contains the following columns:

group a numeric vector. Group codes are 1: Exposed > 10 years, 3: Not exposed.
age a numeric vector. Age in years.
vital.capacity a numeric vector. Vital capacity (a measure of lung volume) in liters.

Source

P. Armitage and G. Berry (1987), *Statistical Methods in Medical Research,* 2nd ed., Blackwell, p. 286.

Examples

```
data(vitcap)
plot(vital.capacity~age, pch=group, data=vitcap)
```

---

vitcap2   *Vital capacity, full data set*

Description

The vitcap2 data frame has 84 rows and 3 columns, age and vital capacity for workers in the cadmium industry.

Format

This data frame contains the following columns:

group a numeric vector. Group codes are 1: Exposed > 10 years, 2: Exposed < 10 years, 3: Not exposed.
age a numeric vector. Age in years.
vital.capacity a numeric vector. Vital capacity (a measure of lung volume) (l).

Source

P. Armitage and G. Berry (1987), *Statistical Methods in Medical Research,* 2nd ed., Blackwell, p. 286.

Examples

```
data(vitcap2)
plot(vital.capacity~age, pch=group, data=vitcap2)
```

---

wright	*Comparison of Wright peak-flow meters*

Description

The wright data frame has 17 rows and 2 columns. It contains data
on peak expiratory flow rate with two different flow meters on each
of 17 subjects.

Format

This data frame contains the following columns:

std.wright a numeric vector. Data from large flow meter (l/min).
mini.wright a numeric vector. Data from mini flow meter (l/min).

Source

J.M. Bland and D.G. Altman (1986), *Lancet*, pp. 307–310.

Examples

```
data(wright)
plot(wright)
abline(0,1)
```

---

zelazo	*Age at walking*

Description

The zelazo object is a list with four components.

Usage

```
data(zelazo)
```

Format

This is a list containing data on age at walking (in months) for four groups of infants:

active test group receiving active training. These children had their walking and placing reflexes trained during four three-minute sessions that took place every day from their second to their eighth week of life.

passive passive training group. Received the same types of social and gross motor stimulation, but did not have their specific walking and placing reflexes trained.

none no training. Had no special training, but were tested along with the children who underwent active or passive training.

ctr.8w eigth-week controls. Had no training and were only tested at the age of 8 weeks.

Note

When asked to enter these data from a text source, many students will use one vector per group and will need to reformat data into a data frame for some uses. The rather unusual format of this data set mimics that situation.

Source

P.R. Zelazo, N.A. Zelazo, and S. Kolb (1972), "Walking" in the newborn, *Science*, 176, 314–315.

Examples

```
data(zelazo)
```

C
Compendium

Elementary

Commands

`ls()` or `objects()`	List objects in workspace
`rm(object)`	Delete `object`
`search()`	Search path

Variable names

Combinations of letters, digits and period. Must not start with a digit. Avoid starting with period.

Assignments

`<-`	Assign value to variable
`->`	Assignment "to the right"
`<<-`	Global assignment (in functions)

Operators

Arithmetic

+	Addition
−	Subtraction, sign
*	Multiplication
/	Division
^	Raise to power
%/%	Integer division
%%	Remainder from integer division

Logical and relational

==	Equal to
!=	Not equal to
<	Less than
>	Greater than
<=	Less than or equal to
>=	Greater than or equal to
is.na(x)	Missing?
&	Logical AND
|	Logical OR
!	Logical NOT

& and | are *elementwise*. See "Programming" (p. 258) for && and ||.

Vectors and data types

Generating

`numeric(25)`	25 zeros
`character(25)`	25 × " "
`logical(25)`	25 × FALSE
`seq(-4,4,0.1)`	Sequence: −4.0 −3.9 3.8 ... 3.9 4.0
`1:10`	Same as `seq(1,10,1)`
`c(5,7,9,13,1:5)`	Concatenation: 5 7 9 13 1 2 3 4 5
`rep(1,10)`	1 1 1 1 1 1 1 1 1 1
`gl(3,2,12)`	Factor with 3 levels, repeat each level in blocks of 2, up to length 12 (i.e., 1 1 2 2 3 3 1 1 2 2 3 3)

Coercion

`as.numeric(x)`	Convert to numeric
`as.character(x)`	Convert to text string
`as.logical(x)`	Convert to logical
`factor(x)`	Create factor from vector x

Re. factors, see also "Tabulation, grouping, recoding" (p. 253).

Data frames

`data.frame(height = c(165,185), weight = c(90,65))`	Data frame with two named vectors
`data.frame(height, weight)`	Collect vectors into data frame
`dfr$var`	Select vector var in data frame dfr
`attach(dfr)`	Put data frame in search path
`detach()`	— and remove it from path

Attached data frames always come *after* `.GlobalEnv` in the search path.

Attached data frames are copies; subsequent changes to `dfr` have no effect.

Numerical functions

Mathematical

`log(x)`	Logarithm of x, natural (base-e) logarithm
`log10(x)`	Base-10 logarithm
`exp(x)`	Exponential function e^x
`sin(x)`	Sine
`cos(x)`	Cosine
`tan(x)`	Tangent
`asin(x)`	Arcsin (inverse sine)
`acos(x)`	
`atan(x)`	
`min(x)`	Smallest value in vector
`min(x1,x2,...)`	minimum over several vectors (one number)
`max(x)`	Largest value in vector
`range(x)`	Like `c(min(x),max(x))`
`pmin(x1,x2,...)`	*Parallel* (elementwise) minimum over multiple equally long vectors
`pmax(x1,x2,...)`	Parallel maximum
`length(x)`	Number of elements in vector
`sum(complete.cases(x))`	Number of non-missing elements in vector

Statistical

`mean(x)`	Average
`sd(x)`	Standard deviation
`var(x)`	Variance
`median(x)`	Median
`quantile(x,p)`	Quantiles
`cor(x,y)`	Correlation

Indexing/selection

`x[1]`	First element
`x[1:5]`	Subvector containing first five elements
`x[c(2,3,5,7,11)]`	Element nos. 2, 3, 5, 7, and 11
`x[y<=30]`	Selection by logical expression
`x[sex=="male"]`	Selection by factor variable
`i <- c(2,3,5,7,11); x[i]`	Selection by numerical variable
`l <- (y<=30); x[l]`	Selection by logical variable

Matrices and data frames

`m[4,]`	Fourth row
`m[,3]`	Third column
`dfr[dfr$var<=30,]`	Partial data frame
`subset(dfr,var<=30)`	Same, often simpler

Input of data

`data(name)`	Built-in data set
`read.table("filename")`	Read from external file

Common arguments to `read.table`

`header=TRUE`	First line has variable names
`sep=","`	Data are separated by commas
`dec=","`	Decimal point is comma
`na.strings="."`	Missing value is dot

Variants of `read.table`

`read.csv("filename")`	Comma separated
`read.delim("filename")`	Tab delimited
`read.csv2("filename")`	Semicolon separated, comma decimal point
`read.delim2("filename")`	Tab delimited, comma decimal point

These all set `header=TRUE`

Missing values

Functions

is.na(x)	Logical vector. TRUE where x has NA
complete.cases(x1,x2,...)	Neither missing in x1, nor x2, nor...

Arguments to other functions

na.rm=	In statistical functions. Remove missing if TRUE, return NA if FALSE.
na.last=	In sort TRUE, FALSE and NA means, respectively, "last", "first", and "throw away".
na.action=	In lm, etc., values na.fail, na.omit, na.exclude. Also in options("na.action").
na.print=	In summary and print.default: How to represent NA in output.
na.strings=	In read.table(): Code(s) for NA in input.

Tabulation, grouping, recoding

`table(f1,...)`	(Cross)tabulation
`tapply(x,f,mean)`	Table of means
`factor(x)`	Convert vector to factor
`cut(x,breaks)`	Groups from cutpoints for continuous variable

Arguments to `factor`

`levels`	Values of x to code. Use if some values are not present in data, or if the order would be wrong.
`labels`	Values associated with factor levels.
`exclude`	Values to exclude. Default NA. Set to NULL to have missing values included as a level.

Arguments to `cut`

`breaks`	Cutpoints. Note values of x outside `breaks` gives NA. Can also be a single number, the number of cutpoints (not very useful).
`labels`	Names for groups. Default is `(0,30]`, etc.
`right`	Right endpoint included? (FALSE: left)

Recoding factors

`levels(f) <- names`	New level names
`factor(newcodes[f])`	Combining levels: `newcodes`, e.g., `c(1,1,1,2,3)` to amalgamate the first 3 of 5 groups.

Statistical distributions

Normal distribution

dnorm(x)	Density
pnorm(x)	Cumulative distribution function, $P(X \le x)$
qnorm(p)	p-quantile, $x : P(X \le x) = p$
rnorm(n)	n (pseudo-)random normally distributed numbers

Distributions

pnorm(x,mean,sd)	Normal
plnorm(x,mean,sd)	Lognormal
pt(x,df)	Student's t
pf(x,n1,n2)	F distribution
pchisq(x,df)	χ^2
pbinom(x,n,p)	Binomial
ppois(x,lambda)	Poisson
punif(x,min,max)	Uniform
pexp(x,rate)	Exponential
pgamma(x,shape,scale)	Gamma
pbeta(x,a,b)	Beta

Same convention (d-q-r) for density, quantiles, and random numbers as for normal distribution.

Statistical standard methods

Continuous response

t.test	One- and two-sample t test
pairwise.t.test	Pairwise comparisons
cor.test	Correlation
var.test	Comparison of two variances (F test)
lm(y ~ x)	Regression analysis
lm(y ~ f)	One-way analysis of variance
lm(y ~ f1 + f2)	Two-way analysis of variance
lm(y ~ f + x)	Analysis of covariance
lm(y ~ x1 + x2 + x3)	Multiple regression analysis
bartlett.test	Bartlett's test (k variances)
Nonparametric:	
wilcox.test	One- and two-sample Wilcoxon test
kruskal.test	Kruskal–Wallis test
friedman.test	Friedman's two-way analysis of variance
cor.test variants:	
method="kendall"	Kendall's τ
method="spearman"	Spearman's ρ

Discrete response

binom.test	Binomial test (incl. sign test)
prop.test	Comparison of proportions
prop.trend.test	Test for trend in relative proportions
fisher.test	Exact test in small tables
chisq.test	χ^2 test
glm(y ~ x1+x2+x3, binomial)	Logistic regression

Models

Model formulas

~	Described by
+	Additive effects
:	Interaction
*	Main effects + interaction
	(a*b = a + b + a:b)
-1	Remove intercept

Classifications are represented by descriptive variable being a *factor*.

Linear and generalized linear models

`lm.out <- lm(y ~ x)`	Fit model and save result
`summary(lm.out)`	Coefficients, etc.
`anova(lm.out)`	Analysis of variance table
`fitted(lm.out)`	Fitted values
`resid(lm.out)`	Residuals
`predict(lm.out, newdata)`	Predictions for new data frame
`glm(y ~ x, binomial)`	Logistic regression

Diagnostics

`rstudent(lm.out)`	Studentized residuals
`dfbetas(lm.out)`	Change in β if obs. removed
`dffits(lm.out)`	Change in fit if obs. removed

Survival analysis

`S <- Surv(time, ev)`	Create survival object
`survfit(S)`	Kaplan–Meier estimate
`plot(survfit(S))`	Survival curve
`survdiff(S ~ g)`	(Log-rank) test for equal survival curves
`coxph(S ~ x1 + x2)`	Cox's proportional hazards model

Graphics

Standard plots

plot()	Scatterplot (and more)
hist()	Histogram
boxplot()	Box-and-whiskers plot
stripplot()	Stripplot
barplot()	Bar diagram
dotplot()	Dot diagram
piechart()	Cakes...
interaction.plot()	Interaction plot

Plotting elements

lines()	Lines
abline()	Line given by intercept and slope (and more)
points()	Points
segments()	Line segments
arrows()	Arrows (NB: angle=90 for error bars)
axis()	Axis
box()	Frame around plot
title()	Title (above plot)
text()	Text in plot
mtext()	Text in margin
legend()	List of symbols

These are all *added* to existing plots.

Graphical parameters

pch	Symbol (*p*lotting *ch*aracter)
mfrow, mfcol	Several plots on one (*m*ulti*f*rame)
xlim, ylim	Plot limits
lty, lwd	Line type/width
col	Colour
cex, mex	Character size and line spacing in margins

See the help page for par for more details.

Programming

| Conditional execution | ```r
if(p<0.05)
    print("Hooray!")
``` |
| — with alternative | ```r
if(p<0.05)
    print("Hooray!")
else
    print("Bah.")
``` |
| Loop over list | ```r
for(i in 1:10)
    print(i)
``` |
| Loop | ```r
i <- 1
while(i<10) {
    print(i)
    i <- i + 1
}
``` |
| User-defined function | ```r
f <- function(a,b,doit=FALSE){
    if (doit)
        a + b
    else
        0
}
``` |
| In flow control one uses a `&&` b and a `||` b where b is only computed if necessary; that is, `if a then b else FALSE` and `if a then TRUE else b` | |
```

Bibliography

Agresti, A. (1990), *Categorical Data Analysis*, John Wiley & Sons, New York.

Altman, D. G. (1991), *Practical Statistics for Medical Research*, Chapman & Hall, London.

Andersen, P. K., Borgan, Ø., Gill, R. D., and Keiding, N. (1991), *Statistical Models Based on Counting Processes*, Springer-Verlag, New York.

Armitage, P. and Berry, G. (1994), *Statistical Methods in Medical Research*, 3rd ed., Blackwell, Oxford.

Becker, R. A., Chambers, J. M., and Wilks, A. R. (1988), *The NEW S Language*, Chapman & Hall, London.

Campbell, M. J. and Machin, D. (1993), *Medical Statistics. A Commonsense Approach*, 2nd ed., John Wiley & Sons, Chichester.

Chambers, J. M. and Hastie, T. J. (1992), *Statistical Models in S*, Chapman & Hall, London.

Clayton, D. and Hills, M. (1993), *Statistical Models in Epidemiology*, Oxford University Press, Oxford.

Cleveland, W. S. (1994), *The Elements of Graphing Data*, Hobart Press, New Jersey.

Cochran, W. G. and Cox, G. M. (1957), *Experimental Designs*, 2nd ed., John Wiley & Sons, New York.

Cox, D. R. (1970), *Analysis of Binary Data*, Chapman & Hall, London.

Cox, D. R. and Oakes, D. (1984), *Analysis of Survival Data*, Chapman & Hall, London.

Everitt, B. S. (1994), *A Handbook of Statistical Analyses Using S-PLUS*, Chapman & Hall, London.

Hájek, J., Šidák, Z., and Sen, P. K. (1999), *Theory of Rank Tests*, 2nd ed., Academic Press, San Diego.

Hald, A. (1952), *Statistical Theory with Engineering Applications*, John Wiley & Sons, New York.

Hosmer, D. W. and Lemeshow, S. (2000), *Applied Logistic Regression*, 2nd ed., John Wiley & Sons, New York.

Johnson, R. A. (1994), *Miller & Freund's Probability & Statistics for Engineers*, 5th ed., Prentice-Hall, Englewood Cliffs, NJ.

Kalbfleisch, J. D. and Prentice, R. L. (1980), *The Statistical Analysis of Failure Time Data*, John Wiley & Sons, New York.

Krause, A. and Olson, M. (1997), *The Basics of S and S-PLUS*, Springer-Verlag, New York.

Lehmann, E. L. (1975), *Nonparametrics, Statistical Methods Based on Ranks*, McGraw-Hill, New York.

Matthews, D. E. and Farewell, V. T. (1988), *Using and Understanding Medical Statistics*, 2nd ed., Karger, Basel.

McCullagh, P. and Nelder, J. A. (1989), *Generalized Linear Models*, 2nd ed., Chapman & Hall, London.

Siegel, S. (1956), *Nonparametric Statistics for the Behavioral Sciences*, McGraw-Hill International, Auckland.

Spector, P. (1994), *An Introduction to S and S-Plus*, Duxbury, Belmont, CA.

Venables, W. N. and Ripley, B. D. (2000), *S Programming*, Springer-Verlag, New York.

Venables, W. N. and Ripley, B. D. (2002), *Modern Applied Statistics with S*, 4th ed., Springer-Verlag, New York.

Weisberg, S. (1985), *Applied Linear Regression*, 2nd ed., John Wiley & Sons, New York.

Zar, J. H. (1999), *Biostatistical Analysis*, Prentice Hall, Englewood Cliffs, NJ.

Index

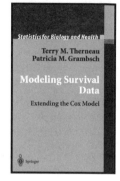